Dinoflagellate Biology: Using Molecular Approaches to Unlock Their Ecology and Evolution

Dinoflagellate Biology: Using Molecular Approaches to Unlock Their Ecology and Evolution

Editor

Shauna Murray

MDPI • Basel • Beijing • Wuhan • Barcelona • Belgrade • Manchester • Tokyo • Cluj • Tianjin

Editor
Shauna Murray
School of Life Sciences
University of Technology Sydney
Sydney
Australia

Editorial Office
MDPI
St. Alban-Anlage 66
4052 Basel, Switzerland

This is a reprint of articles from the Special Issue published online in the open access journal *Microorganisms* (ISSN 2076-2607) (available at: www.mdpi.com/journal/microorganisms/special_issues/dinoflagellate).

For citation purposes, cite each article independently as indicated on the article page online and as indicated below:

LastName, A.A.; LastName, B.B.; LastName, C.C. Article Title. *Journal Name* **Year**, *Volume Number*, Page Range.

ISBN 978-3-0365-6198-1 (Hbk)
ISBN 978-3-0365-6197-4 (PDF)

Cover image courtesy of Shauna Murray

Contents

About the Editor

Shauna Murray

Professor Shauna Murray is a marine biologist whose research focuses on ecology and evolution, using molecular genetic techniques. Marine biotoxins produced by dinoflagellates, single-celled marine protists, are amongst the most toxic substances recorded to date; however, we know little about the genetics, evolution, and ecology of these organisms. Professor Murray is an expert on the systematics, genetics, evolution, and toxins of marine dinoflagellates, with a particular focus on benthic and harmful algal species. She works extensively with the aquaculture and fisheries sectors and is based at the University of Technology Sydney in Australia.

 *microorganisms*

MDPI

Article

Improved *Cladocopium goreaui* Genome Assembly Reveals Features of a Facultative Coral Symbiont and the Complex Evolutionary History of Dinoflagellate Genes

Yibi Chen [1], Sarah Shah [1], Katherine E. Dougan [1], Madeleine J. H. van Oppen [2,3], Debashish Bhattacharya [4] and Cheong Xin Chan [1,*]

1 Australian Centre for Ecogenomics, School of Chemistry and Molecular Biosciences, The University of Queensland, Brisbane, QLD 4072, Australia
2 School of Bioscience, The University of Melbourne, Parkville, VIC 3010, Australia
3 Australian Institute of Marine Science, Townsville, QLD 4810, Australia
4 Department of Biochemistry and Microbiology, Rutgers University, New Brunswick, NJ 08901, USA
* Correspondence: c.chan1@uq.edu.au

Abstract: Dinoflagellates of the family Symbiodiniaceae are crucial photosymbionts in corals and other marine organisms. Of these, *Cladocopium goreaui* is one of the most dominant symbiont species in the Indo-Pacific. Here, we present an improved genome assembly of *C. goreaui* combining new long-read sequence data with previously generated short-read data. Incorporating new full-length transcripts to guide gene prediction, the *C. goreaui* genome (1.2 Gb) exhibits a high extent of completeness (82.4% based on BUSCO protein recovery) and better resolution of repetitive sequence regions; 45,322 gene models were predicted, and 327 putative, topologically associated domains of the chromosomes were identified. Comparison with other Symbiodiniaceae genomes revealed a prevalence of repeats and duplicated genes in *C. goreaui*, and lineage-specific genes indicating functional innovation. Incorporating 2,841,408 protein sequences from 96 taxonomically diverse eukaryotes and representative prokaryotes in a phylogenomic approach, we assessed the evolutionary history of *C. goreaui* genes. Of the 5246 phylogenetic trees inferred from homologous protein sets containing two or more phyla, 35–36% have putatively originated via horizontal gene transfer (HGT), predominantly (19–23%) via an ancestral Archaeplastida lineage implicated in the endosymbiotic origin of plastids: 10–11% are of green algal origin, including genes encoding photosynthetic functions. Our results demonstrate the utility of long-read sequence data in resolving structural features of a dinoflagellate genome, and highlight how genetic transfer has shaped genome evolution of a facultative symbiont, and more broadly of dinoflagellates.

Keywords: dinoflagellates; genome; *Cladocopium goreaui*; phylogenomics; horizontal gene transfer

Citation: Chen, Y.; Shah, S.; Dougan, K.E.; van Oppen, M.J.H.; Bhattacharya, D.; Chan, C.X. Improved *Cladocopium goreaui* Genome Assembly Reveals Features of a Facultative Coral Symbiont and the Complex Evolutionary History of Dinoflagellate Genes. *Microorganisms* **2022**, *10*, 1662. https://doi.org/10.3390/microorganisms10081662

Academic Editor: Shauna Murray and Evgene Rosenberg

Received: 1 July 2022
Accepted: 15 August 2022
Published: 17 August 2022

Publisher's Note: MDPI stays neutral with regard to jurisdictional claims in published maps and institutional affiliations.

1. Introduction

Dinoflagellates of the family Symbiodiniaceae [1] are diverse microalgae, with many forming symbiotic relationships that are critical to corals and other coral reef organisms. Symbiodiniaceae provide carbon fixed via photosynthesis and other essential nutrients to coral hosts [2,3]. Environmental stress leads to breakdown of this partnership and loss of the algae, i.e., coral bleaching, putting the corals at risk of starvation, disease, and potential death [4]. Recent studies of Symbiodiniaceae genomes have revealed extensive sequence and structural divergence [5–8], and potentially a greater, yet-to-be recognised phylogenetic diversity among these taxa [9,10]. A recent comparative analysis of genomes from 18 dinoflagellate taxa (of which 16 are Symbiodiniaceae) revealed distinct phylogenetic signals between genic and non-genic regions [11], indicating differential evolutionary pressures acting on these genomes. These findings illustrate how the evolutionary complexity of Symbiodiniaceae genomes may explain their diverse symbioses and ecological niches [12].

Cladocopium (formerly Clade C) is the most taxonomically diverse genus of family Symbiodiniaceae and found predominantly in the Indo-Pacific, with *Cladocopium goreaui* (formally type C1) being a dominant species in the region [13,14]. The earlier genome analysis of *C. goreaui* SCF055 [7] revealed the genetic capacity of the species to establish and maintain symbiosis with coral hosts, respond to stress, and to undergo meiosis; i.e., many of the implicated genes show evidence of positive selection. Although these results provide insights into the adaptive evolution of genes, the assembled genome, generated using only Illumina short-read data, remains fragmented with 41,289 scaffolds [7]. Additional analysis of the draft genome also indicated that some scaffolds may be of bacterial origin due to their anomalous G+C content [15]. For these reasons, the existing data limit our capacity to reliably assess repetitive genomic elements and evolutionary origins of the predicted genes.

Here, we present an improved, hybrid genome assembly for *C. goreaui*, combining novel PacBio long-read sequence data with the existing short-read sequence data from Liu et al. [7], and incorporating a new full-length transcriptome to guide gene prediction. Incorporating proteins predicted from the genome with those from 96 taxonomically broadly sampled eukaryote taxa and representative prokaryotes in a phylogenomic analysis, we assessed the evolutionary origins of genes in *C. goreaui* and other dinoflagellates, and the impact of horizontal genetic transfer (HGT) in shaping the evolution of this lineage. The earlier investigation based on transcriptome data from the bloom-forming, toxin-producing dinoflagellate *Alexandrium tamarense* [16] revealed evidence of HGT, implicating both prokaryote and eukaryote donors, in 14–17% of investigated protein trees. Few genes and no genome data were available from other dinoflagellates when that study was conducted. However, the genomic and genetic resources of dinoflagellates have grown appreciably in the last decade, enabling a more-balanced representation of taxonomic diversity to support such an analysis. In this study, we examine how the nuclear genome of *C. goreaui*, and broadly that of dinoflagellates, has evolved and benefited from the acquisition of genomic (and functional) novelties through HGT.

2. Results and Discussion

2.1. Improved C. goreaui Genome Assembly Reveals More Repeats and More Duplicated Genes

We generated PacBio long-read genome data (50.2 Gbp; Table S1) for *C. goreaui* SCF055 and combined them with existing Illumina short reads in a hybrid approach to generate a de novo genome assembly (see Methods). The first published genome assembly of the SCF055 isolate [7] was previously refined to exclude putative contaminant sequences [15]. Compared to the draft assembly reported in Chen et al. [15], our assembly exhibits a five-fold decrease in the number of scaffolds (6843) and a three-fold increase in scaffold N50 length (354 Kbp; Table 1). Genome size was estimated at 1.3 Gbp based on *k*-mers (Figure S1), and our improved assembly (1.2 Gbp in size) is larger than the earlier draft (1.0 Gbp; Table 1).

We also generated 65,432 near-full-length transcripts using PacBio Iso-Seq to guide prediction of protein-coding genes. Using the same approach tailored for dinoflagellates [15], we predicted 45,322 protein-coding genes (mean length 15,745 bp) in the genome, compared to 39,066 (mean length 8428 bp) reported in Chen et al. [15]. The majority (82.4%) of predicted genes are supported by transcript evidence, and genome completeness is markedly improved, evidenced by the 15.1% greater recovery of core conserved genes (BUSCO alveolate_odb10) [17] at 82.4% (Table 1). Most predicted proteins (40,495; 89.3%) have UniProt hits based on sequence similarity (BLASTp; $E \leq 10^{-5}$); 19,904 (43.9%) have hits in the curated Swiss-Prot database, 8836 (19.5%) covering > 90% of full-length Swiss-Prot proteins. The remaining 4827 (10.7%) *C. goreaui* proteins have no significant hits in UniProt, indicating the prevalence of "dark" genes that encode functions yet to be discovered.

Table 1. Metrics of the revised genome assembly and predicted genes of *C. goreaui*, compared to the earlier assembled genome.

Metric		Earlier Assembly [15]	Revised Assembly (This Study)
Assembly size (Gbp)		1.0	1.2
Number of scaffolds		41,235	6843
Genome scaffolds N50 (Kbp)		91	354
Genome GC-content (%)		44.76	44.38
Number of predicted genes		39,006	45,322
Percentage of BUSCO proteins recovered (alveolata_odb10)		67.3	82.4
Genes with transcript support (%)		76.5	82.5
Average gene length (bp)		8428	15,745
Average CDS length (bp)		1625	2018
Total gene length (Mbp)		328.7	713.6
Total CDS length (Mbp)		63.4	91.5
Average number of exons per gene		12.4	17.2
Average exon length (bp)		130.4	120.4
Genes with introns (%)		95.9	95.9
Number of introns per gene		11.4	16.2
Average intron length (bp)		593.5	838.8
Splice donor motif (%)	GT	35.7	36.6
	GC	43.3	43.6
	GA	20.8	19.8
Splice acceptor with AGG motif (%)		96.3	96.1
Number of intergenic regions		24,243	39,720
Average length of intergenic regions (bp)		9539	7388

We identified and compared repeat content in the *C. goreaui* genome with that in the earlier assembly of Chen et al. [15]. Excluding simple repeats, we found a higher repeat content (36.5% of total bases in the assembled genome; Figure 1a) in the current assembly than in the initial data (21.1%), with known repeat types accounting for 17.3% of total bases, compared to 5.6%. This result indicates a better resolution of repetitive regions in the revised genome with the incorporation of long reads. Novel, Symbiodiniaceae-specific repeats would remain as unclassified in this instance due to the lack of identified dinoflagellate repeats in the current database. Among known repeat types, long terminal repeats (LTR) and DNA transposons are the most prevalent repeat families (constituting 7.3% and 6.2% of total bases, respectively). These two repeat families exhibit distinct levels of sequence divergence; those with Kimura substitution values centred between 0 and 10 are more conserved than those with values centred between 15 and 25. Most LTRs are highly conserved in *C. goreaui*, a trend also observed in the genomes of other dinoflagellates [10,18].

Collinear gene blocks within a genome represent duplicated gene blocks, e.g., via segmental or whole-genome duplication. Based on the recovery of these gene blocks, we identified a greater proportion of duplicated genes than Chen et al. [15]: 35,119 (77.5% of 45,322) genes in duplicates, compared to 25,550 (65.5% of 39,006; Table 2). We found 31,827 (70.2%) genes in dispersed duplicates, suggesting a lack of conserved collinearity of genes in the *C. goreaui* genome. The lack of collinearity of duplicated genes lends support to the hypothesized extensive structural rearrangements in genomes of facultative symbionts in the family Symbiodiniaceae [12]; more genome data from non-facultative or free-living taxa will allow for a more-robust testing of this hypothesis. With the improved structural annotation, we recovered a 2.39-fold greater number of tandemly repeated genes, and 387 genes (in 34 collinear blocks) implicated in segmental duplications (Table 2). Tandemly repeated genes in dinoflagellates are thought to be a mechanism for improving transcriptional efficiency, particularly for genes encoding critical functions [18]. Comparing gene ontology (GO) terms annotated in the 1998 tandemly repeated genes against those in all *C. goreaui* genes, the top three enriched terms for Cellular Component (Table S2) are "chloroplast thylakoid membrane" (GO:0009535; $p = 1.5 \times 10^{-7}$), "photosystem I reaction center" (GO:0009538; $p = 5.4 \times 10^{-7}$) and "photosystem II" (GO:0009523; $p = 0.00042$). This result indicates that genes encoding photosynthetic functions tend to appear in tandem repeats in *C. goreaui*. We applied the same approach to genes in segmental duplications and found that the most significantly enriched Biological Process term for this set is "recombinational repair" (GO:0000725, $p = 7.2 \times 10^{-5}$, Table S3). The repair of errors during genetic recombination is essential for maintaining genome integrity; conservation of collinear organization of these genes may be key in ensuring their transcription efficiency.

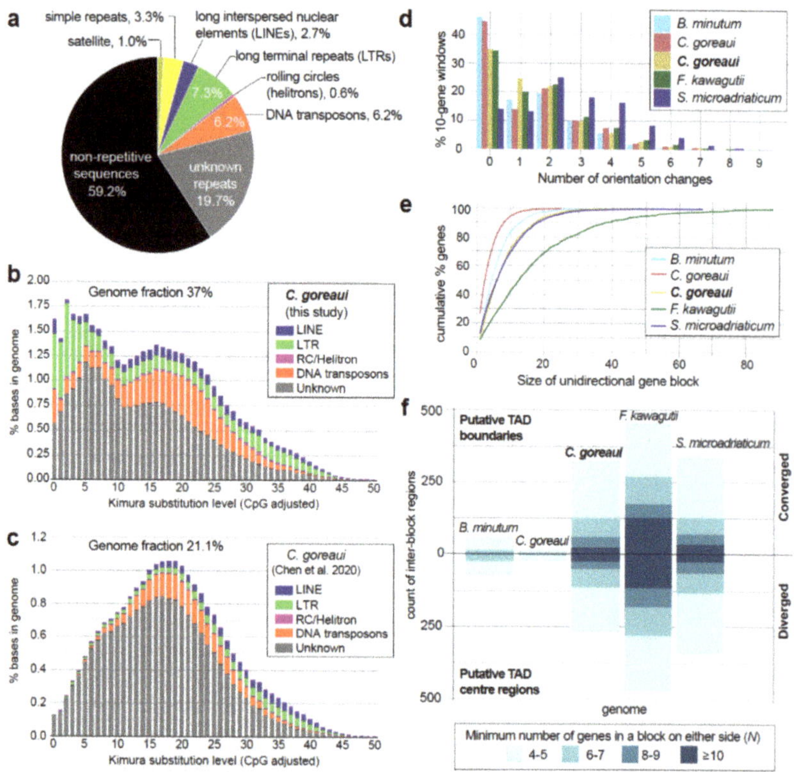

Figure 1. Genome features of *C. goreaui*. (**a**) Repeat families identified in the revised genome assembly; repeat landscape shown for the (**b**) revised genome assembly of *C. goreaui* (highlighted in bold-face) and (**c**) the earlier assembly of Chen et al. [15]. (**d**) Frequency of strand-orientation change in 10-gene windows and (**e**) cumulative percentage of genes in unidirectionally encoded blocks, shown for five representative Symbiodiniaceae genomes: *Breviolum minutum* [15], *C. goreaui* [15], *C. goreaui* (boldfaced, this study), *Fugacium kawagutii* [19], and *Symbiodinium microadriaticum* [20]; and (**f**) number of inter-block regions in each genome assembly indicating putative TAD central regions and boundaries, shown for representative genomes, based on the minimum number of genes in a unidirectional block. Bars above the *x*-axis represent inter-block regions at which orientations of unidirectional blocks converged, whereas bars below the *x*-axis represent those at which the orientations diverged.

Table 2. Types of gene duplication identified in *C. goreaui*.

Duplication Type	Earlier Assembly [15]	Revised Assembly (This Study)
Singleton	13,456 (34.5%)	10,203 (22.5%)
Dispersed	24,441 (62.6%)	31,827 (70.2%)
Proximal	273 (0.7%)	907 (2.0%)
Tandem	836 (2.1%)	1998 (4.4%)
Segmental	0 (0.0%)	387 (0.9%)

2.2. Topologically Associated Domains (TADs) and Unidirectional Gene Blocks

Recent studies have clarified interacting genomic regions via topologically associated domains (TADs) of dinoflagellate chromosomes [20,21]. Orientations of unidirectionally encoded gene blocks diverge from a TAD central region, converging at TAD boundaries [20]. Regulatory elements such as promoters and enhancers of gene expression are hypothesised to be concentrated in these regions to regulate transcription of the upstream or downstream unidirectional gene blocks [22]. To assess putative TAD regions in the revised *C. goreaui* genome, we first identified the unidirectionally encoded gene blocks. We followed the method of Stephens et al. [18] to enumerate gene-orientation change(s) in a ten-gene window, moving across the entire genome assembly; the tendency for no change in gene orientation is an indication for the prevalence of unidirectional encoding. We performed this analysis in a set of representative Symbiodiniaceae genomes (Figure 1d). Interestingly, we observed a lower percentage (34.7%) of ten-gene windows with conserved orientation in the revised *C. goreaui* genome, when compared to 44.6% in the assembly of Chen et al. [15]. The equivalent percentages in the more-contiguous, near-chromosomal level genome assemblies of *S. microadriaticum* [20] (13.9%) and *F. kawagutii* [19] (34.5%) are also lower, compared to 46.1% in the more-fragmented assembly of *B. minutum* [15] (Figure 1d). However, when we assessed the sizes of unidirectional gene blocks in these genomes, they are clearly larger in the more-contiguous assemblies (Figure 1e). For instance, 32.6% of genes in *C. goreaui* are found in block sizes of 10 or more genes, compared to only 7.6% in the earlier assembly (Figure 1e). These observations indicate that the lower recovery of ten-gene windows with conserved orientation is caused by the increased recovery of windows spanning two gene blocks with opposing orientations, as expected in TAD central or boundary regions, in the more-contiguous assemblies. In this way, the more-contiguous assembly enables better recovery of putative TAD regions.

To assess putative TAD regions, we examined genomic regions between any two unidirectional gene blocks. We identified these regions requiring the gene blocks on either side to contain at least N number of genes, where N is 4, 6, 8, or 10. Figure 1f shows the recovery of these regions across threshold N in the representative genomes, with those with converging gene-block orientations (i.e., putative TAD boundaries) above the x-axis, and those with diverging orientations (i.e., putative TAD central regions) below the x-axis. We recovered approximately 6- to 30-fold larger numbers of these regions in the more-contiguous *C. goreaui* assembly (e.g., 327 putative TAD boundaries) and in near-chromosomal level assemblies of *S. microadriaticum* (340) and *F. kawagutii* (454), than in the more-fragmented assemblies of *B. minutum* (59) and *C. goreaui* (15). The implicated unidirectional gene blocks on either side of these regions are also larger, e.g., at $N = 10$, we identified 25 putative TAD boundaries in *C. goreaui*, compared to only two in the earlier assembly; the assembly of *F. kawagutii* shows the greatest recovery of TAD-associated regions, with 129 putative TAD boundaries implicating blocks of 10 or more genes on either side. TAD boundaries have been reported to exhibit a dip in GC content in the middle of the sequence [20]. We observed such a dip in GC content in 17/25 putative TAD boundary regions (at $N = 10$) in *C. goreaui*; an example is shown in Figure S2. Interestingly, our recovery of TAD-associated regions in *C. goreaui* is very similar to that in the chromosome-level assembly of *S. microadriaticum* (Figure 1f), suggesting that our revised assembly, although not derived specifically using chromosome configuration capture, e.g., in Nand et al. [20], resolves a comparable number of TAD regions.

We also identified genes that tend to disrupt the unidirectional coding of gene blocks, based on their distinct orientation from upstream and downstream genes; such disruptions have been observed in the chromosome-level genome assembly of *S. microadriaticum* [20]. We identified 3799 (8.4%) of such genes in *C. goreaui*. Interestingly, these genes largely encode functions related to transposon elements. Comparing the annotated GO terms in these genes versus those in all *C. goreaui* genes, the two most significantly enriched terms for Molecular Function are "nucleic acid binding" (GO:0003676; $p < 1.0 \times 10^{-30}$) and "RNA-DNA hybrid ribonuclease activity" (GO:0004523; $p = 1.8 \times 10^{-20}$; Table S4), and the most significantly enriched term for Biological Process is "DNA integration" (GO:0015074; $p = 7.6 \times 10^{-7}$; see Table S4). We found no expression evidence for genes encoding transposable elements that disrupt the orientation of unidirectional gene blocks. In contrast, gene encoding transposable elements that occur in the same orientation within unidirectional blocks were actively expressed. This observation suggests that unidirectional encoding is essential for gene expression in *C. goreaui*, and potentially also in dinoflagellates.

2.3. Evolutionary Origin of C. goreaui Genes

The predicted genes from the improved *C. goreaui* genome assembly provide an excellent analysis platform to assess their evolutionary origins. To assess the overall phylogenetic signal of *C. goreaui* genes relative to other dinoflagellates, we inferred a dinoflagellate phylogeny (Supplementary Figure S3) using 3411 single-copy, strictly orthologous protein sets identified using 1,468,870 sequences from 30 dinoflagellate taxa including *C. goreaui*, plus 22,958 sequences from *Perkinsus marinus* as an outgroup (Table S5; see Methods). This phylogeny is congruent with the established phylogeny of dinoflagellates [23,24] with distinct strongly supported clades (bootstrap support [BS] ≥ 90%). *C. goreaui* is placed in a well-supported (BS 100%) clade of family Symbiodiniaceae, and within the order Suessiales (BS 100%) to which the family belongs. This result confirms the phylogenetic position of *C. goreaui* in the Symbiodiniaceae based on putative orthologous proteins, a result that has been demonstrated recently based on whole-genome sequence data using an alignment-free phylogenetic approach [11].

We then assessed the evolutionary origin of individual *C. goreaui* genes using protein data. Using 2,841,408 predicted protein sequences from 96 taxa of eukaryotes and prokaryotes (Table S5), we identified 177,346 putative homologous proteins sets based on sequence similarity (see Methods). Of the 45,322 *C. goreaui* proteins, 22,026 (48.6%) are specific to dinoflagellates (i.e., 3021, 8748, and 10,257 are specific to *C. goreaui*, order Suessiales, and Dinophyceae, respectively; Figure 2a). Dinophyceae-specific proteins are those found in one or more Dinophyceae taxa that may also include Suessiales. Although we cannot dismiss biased taxon sampling due to the scarcely available data from other alveolate taxa, these results indicate extensive lineage-specific innovation of gene functions following speciation or divergence of dinoflagellates, supporting the notion of extreme divergence of dinoflagellate genes [5,9,10,24].

We found 4601 (10.2% of 45,322) proteins to be shared exclusively with another phylum, in 1544 homologous sets (Figure 2b). Assuming that inadequate sampling is less of a concern in sets that contain a larger number of genes, we adapted the approach by Chan et al. [16] to study these putative gene-sharing partners (phylum) with dinoflagellates, based on the minimum number of genes (x) in each set, at $x \geq 2$, ≥ 20, ≥ 40, and ≥ 60. At $x \geq 2$, the most frequent gene-sharing partners for dinoflagellates are Chromerida, Perkinsea, and other alveolates (534), followed by Stramenopiles (195), Haptophyta (162), and Archaeplastida (146). This result supports the current phylogeny of dinoflagellates in the supergroup Alveolata, the Stramenopiles+Alveolata+Rhizaria (SAR) clade [25–27], and their close association with haptophytes [28] and Archaeplastida via endosymbiosis implicated by their plastid origin [29,30]. The earlier study based on transcriptome data from the bloom-forming dinoflagellate *Alexandrium tamarense* [16] revealed a decrease in the proportion of dinoflagellate genes shared with alveolates as x increased. This trend is not observed here, e.g., the percentage of genes showing exclusive sharing with alveolates is 34.6%, 37.9%,

38.5%, and 37.3% at $x \geq 2$, ≥ 20, ≥ 40, and ≥ 60, respectively. This result suggests that the phylogenetic signal observed here is more consistent than in the earlier study, boosted by the greater representation of dinoflagellate taxonomic diversity (30 taxa in Table S5).

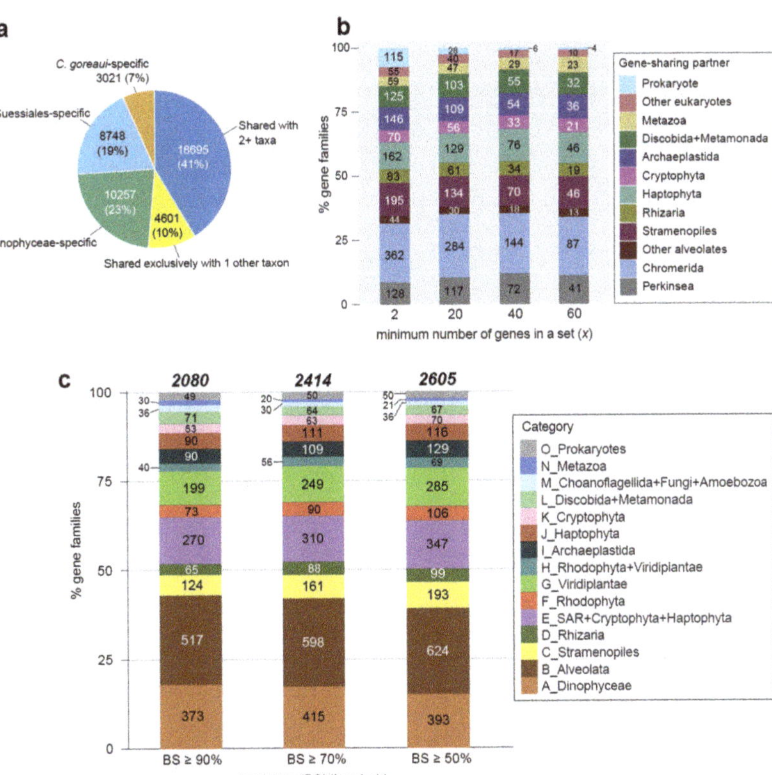

Figure 2. Evolutionary origins of *C. goreaui* genes. (**a**) *C. goreaui* genes classified based on the number of recovered protein homologs in other taxa. (**b**) Distribution of phyla with respect to exclusive gene-sharing partners for *C. goreaui*, based on the number of homologous sets that contain only *C. goreaui* and the other phylum, across the minimum number of genes in each set (x) at $x \geq 2$, ≥ 20, ≥ 40, and ≥ 60. (**c**) Distribution of phyla that are found to share genes with dinoflagellates, based on the number of inferred protein trees in which dinoflagellates and one other phylum were recovered in a monophyletic clade, assessed at bootstrap support (BS) $\geq 90\%$, $\geq 70\%$, and $\geq 50\%$. All taxonomic classification follows NCBI Taxonomy, including Dinophyceae (Fritsch 1927).

The remaining 18,695 (41.2%) *C. goreaui* proteins were recovered in 5795 homologous sets containing two or more phyla. To assess the evolutionary origins of these genes, we inferred a phylogenetic tree for each of these homologous protein sets; 5246 remained after passing the initial composition chi-squared test in IQ-TREE to exclude sequences for which the character composition significantly deviates from the average composition in an alignment (see Methods). Among the 5246 trees, we adopted a computational approach [31] to identify those in which Dinophyceae taxa form a strongly supported clade with one other phylum, based on observed BS at $\geq 90\%$, $\geq 70\%$, and $\geq 50\%$ (Figure 2c; see Methods and Table S6 for detail of our tree-sorting strategy); clades observed at a higher BS threshold represent higher confidence results. All sorted trees using the three thresholds are available as Supplementary Data 1. We identified 2080, 2414, and 2605 trees that fit these criteria at BS $\geq 90\%$, $\geq 70\%$, and $\geq 50\%$ (Figure 2c); the classification of evolutionary origin for each sorting process is shown in Table S6. The proportions of distinct putative origins for the protein sets are similar, e.g., those with putative alveolate origin are 24.9%, 24.8%, and 24.0% at BS $\geq$ 90%, $\geq$ 70%, and $\geq$ 50% (Figure 2c), reflecting the consistent phylogenetic signal we recovered from these data. Remarkably, the evolutionary history of dinoflagellate proteins in more than one-half of the analysed 5246 trees are too complicated to be classified using our approach, e.g., 2641 (50.3%) even at our least-stringent threshold of BS $\geq$ 50%. Some of these proteins (e.g., acyl-CoA dehydrogenase and GTP-binding protein of YchF family) are thought to have a shared origin with fungi or pico-prasinophytes [16], which are likely to be artefacts due to limited dinoflagellate genome data and sampling bias.

2.4. Genes Implicating a History of Horizontal Transfer

Trees containing a strongly supported monophyletic clade of dinoflagellates and taxonomically remote phyla (Categories F through O in Figure 2c) suggest a history of HGT; they account for 35.1%, 34.9%, and 36.4% of sorted trees at BS $\geq$ 90%, $\geq$ 70%, and $\geq$ 50%. The proportion of HGT-implicated trees is greater than that (14–17%) in the earlier study based on the transcriptome of *Alexandrium tamarense* [16]. In this study, a more taxonomically balanced set of eukaryote taxa was used, including a larger representation of dinoflagellates and red algae. Therefore, the biases introduced by poor taxon-sampling are diminished, as demonstrated by the consistent phylogenetic signal that we captured at different stringency levels.

Dinoflagellates possess secondary (and some tertiary) plastids independently acquired from several algal lineages through endosymbioses [30,32]. Genes from the endosymbionts were postulated to have been transferred to the host nuclear genome during this process. The implicated endosymbionts include the ancestral Archaeplastida lineages of red and/or green algae, and potentially other eukaryotic microbes, e.g., haptophytes, allowing genetic transfer between dinoflagellates and these algal lineages during these events [28]. Secondary plastids found in dinoflagellates and diatoms (stramenopiles) are thought to have arisen from an ancestral red alga [33]; both red and green algal derived genes have been described in these taxa [16,34,35].

Here, we focus on the high-confidence trees (clade recovery at BS $\geq$ 90%) as strong evidence for HGT. Of the 2080 trees, 402 (19.3%) putatively derived from Archaeplastida (groups F through I in Figure 2c): 73 from Rhodophyta (F), 199 from Viridiplantae (G), and 130 from any combination of Archaeplastida taxa (H and I). At the less-stringent BS threshold, this number is 589 (22.6% of 2605). As the most plausible explanation, *C. goreaui* (and dinoflagellate) genes in these trees likely arose via endosymbiotic genetic transfer due to one or more plastid endosymbioses, implicating ancestral Archaeplastida phyla, more evidently with green algae (in Viridiplantae) than with the red algae (Rhodophyta).

Figure 3a shows the phylogenetic tree of beta-glucan synthesis-associated protein homologs, with a strongly supported (BS 97%) monophyletic clade containing Viridiplantae (i.e., the green algae *Chlamydomonas*, *Volvox*, and *Chlorella*), haptophytes, Stramenopiles (including diatoms), and dinoflagellates. The beta-glucans are key components of cellulose that form the thecate armour of the cell wall of dinoflagellates [36], as well as key carbohydrate storage [37]. This tree supports a putative green algal origin of the gene associated with beta-glucan synthesis in dinoflagellates and other closely related taxa, implicating an ancient HGT among these lineages. This is a more parsimonious explanation than to invoke massive gene losses in other alveolates and microbial eukaryotes.

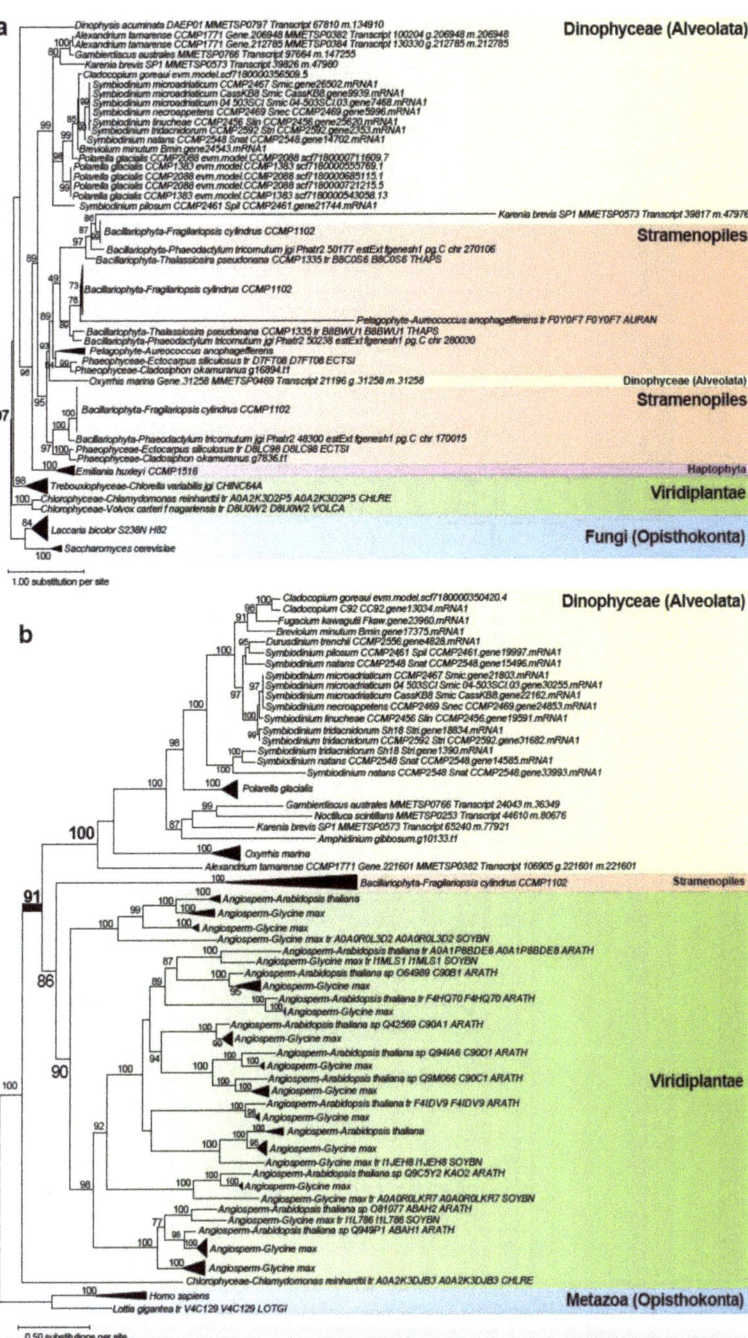

Figure 3. Maximum likelihood trees of (**a**) beta-glucan synthesis-associated protein and (**b**) abscisic acid 8′-hydroxylase, suggesting ancient gene origins from Viridiplantae. The ultrafast bootstrap support of IQ-TREE2 is shown at each internal node; only values ≥ 70% are shown. Unit of branch length is the number of substitutions per site.

Figure 3b shows the tree for putative abscisic acid 8′-hydroxylase, in which Viridiplantae taxa (largely plants) form a strongly supported (BS 100%) monophyletic clade with the diatom *Fragilariopsis cylindrus* and dinoflagellates. This tree supports a Viridiplantae origin of dinoflagellate genes, and subsequent divergence among the Suessiales (BS 98%) including *C. goreaui*. This enzyme is involved in regulating germination and dormancy of plant seeds via oxidation of the hormone abscisic acid [38]. It is also known as cytochrome P450 monooxygenase [39]. In dinoflagellates, this enzyme is known to regulate encystment and maintenance of dormancy [40], and the expression of this gene was found to be upregulated as an initial response to heat stress in a *Cladocopium* species [41]. The tree in Figure 3b shows that the protein homologs in dinoflagellates share sequence similarity to sequences in plants, whereas homologs from other green algae (the more-likely sources of HGT) are absent; given that plants are derived from green algal lineages, this result suggests that the green algal derived gene in plants, dinoflagellates, and the diatom *F. cylindrus* was likely subjected to differential functional divergence or gene loss among the green algae.

We also found evidence for more-recent genetic exchanges. Figure 4 shows the tree for a putative sulphate transporter, which has a strongly supported (BS 93%) monophyletic clade containing dinoflagellates and Viridiplantae (mostly green algae), separate from the usual sister lineage of Stramenopiles expected in the SAR grouping in the eukaryote tree of life [25]. This protein is involved in sulphate uptake, which in green algae has a direct impact on protein biosynthesis in the plastid [42]. This tree suggests a putative green algal origin of the genes in dinoflagellates, which implicates more-recent HGT than those observed in Figure 3. In contrast, some green algal derived genes appear to have undergone duplication upon the diversification of dinoflagellates, e.g., the tree of a hypothetical protein shown in Figure S4. Although the function of this protein in dinoflagellates remains unclear, the homolog in *Arabidopsis thaliana* (UniProt Q94A98; At1g65900) is localised in the chloroplast and implicated in cytokinesis and meiosis [43,44], lending support to an endosymbiotic genetic transfer associated with plastid evolution [45]. Our observation of ancient and recent genetic exchanges between green algae and dinoflagellates suggests HGT is a dynamic and ongoing process in dinoflagellate genome evolution.

Red algal origin of secondary plastids is well established [33,46]. The stronger signal of green algal than red algal origin we observed here based on a taxonomically broad and dinoflagellate-rich dataset (Table S5) lends support to the notion of an additional cryptic green algal endosymbiosis in the evolution of secondary plastids, instead of the "shopping bag" hypothesis that postulates equal proportions of acquired red or green algal genes [47]. Although green algal derived plastids in some dinoflagellates are also known [48,49], these taxa are not included in our analysis here due to a lack of genome data.

We observed a small proportion (7.1%) of trees that suggest putative genetic exchange between dinoflagellates with distantly related eukaryotes and prokaryotes, e.g., groups L through O in Figure 2c. We cannot dismiss that some of these may be artefacts due to sampling bias or even misidentified sequences in the database. For instance, the phylogeny of phosphatidylinositol 4-phosphate 5-kinase (Figure S5) shows a strongly supported (BS 100%) clade containing 55 dinoflagellate sequences and 1 sequence from the coral *Porites lutea* representing Metazoa; this may be a case of misidentification of the sequence from the dinoflagellate symbiont associated with the coral.

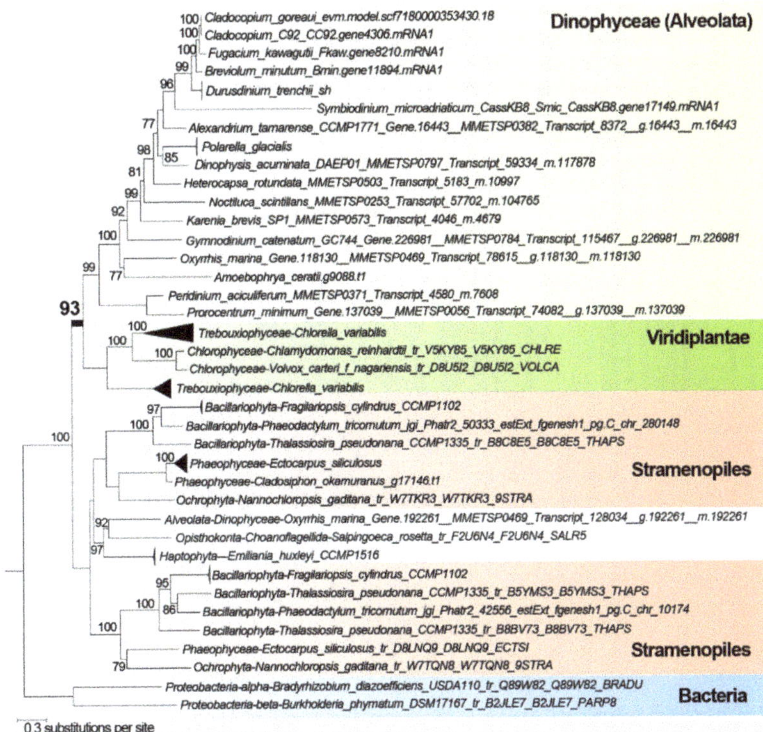

Figure 4. Maximum likelihood tree of putative sulphate transporter indicating a Viridiplantae origin in dinoflagellate genes. The ultrafast bootstrap support of IQ-TREE2 is shown at each internal node; only values ≥ 70% are shown. Unit of branch length is the number of substitutions per site.

2.5. Genes Implicating Vertical Inheritance

Among the high-confidence trees (recovery at BS ≥ 90%), 64.9% (groups A through E in Figure 2c) provide strong evidence of vertical inheritance; these trees contain a strongly supported (BS ≥ 90%) monophyletic clade containing dinoflagellates only (373), and dinoflagellates plus another closely related taxa of Alveolata (517), Stramenopiles (124), Rhizaria (65), and with the SAR group in the presence of Haptophyta and Cryptophyta (270), as expected based on our current understanding of eukaryote tree of life [25]. Figure S6 shows an example of these trees, specifically, ubiquitin carboxyl-terminal hydrolase. In this tree, all the major phyla are mostly well-resolved in strongly supported clades, e.g., Dinophyceae, Rhizaria, Stramenopiles, and Rhodophyta, each at BS 100%, and the clade of Alveolata+Rhizaria (BS 90%). Figure 5 shows another tree that contains a strongly supported (BS 100%) monophyletic clade of the SAR group, within which three clades (two supported at BS 99%, one at BS 100%), each containing similar dinoflagellate taxa, are recovered, highlighting gene-family expansion. Proteins within the three subclades putatively code for distinct functions: (a) an autophagy-related protein 18a (as with other non-dinoflagellate proteins positioned elsewhere in the tree), (b) a transmembrane protein 43, and (c) the pentatricopeptide repeat-containing protein GUN1. These distinct functions were identified, for each sub-clade, based on the annotated function of the top UniProt hit for the sequences within, and their distinct domain configurations (Figure S7). This observation indicates vertical inheritance of the gene encoding transmembrane protein in dinoflagellates, which was then duplicated and underwent neo- or sub-functionalization to generate functional diversity. This result aligns with known features of gene-family evolution that generates functional novelty in dinoflagellates, including tandemly repeated genes that encode adaptive functions [18,50].

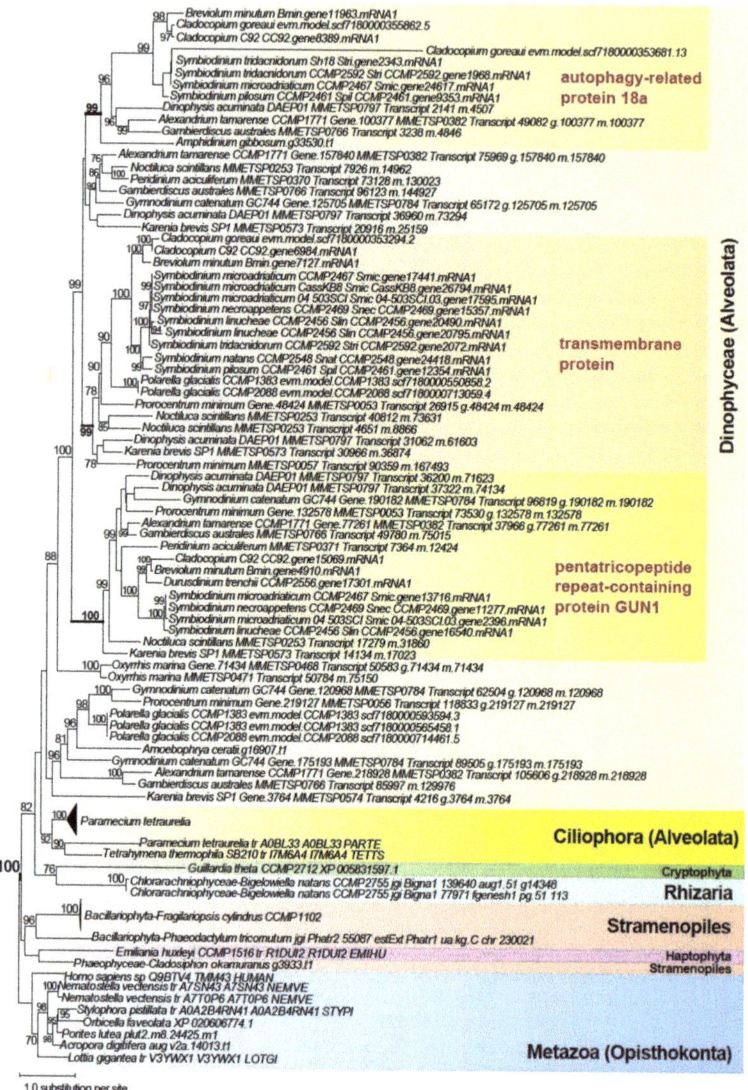

Figure 5. Maximum likelihood protein tree showing vertical inheritance and gene expansion among dinoflagellates, with distinct clades containing the autophagy-related protein 18a, the transmembrane protein 43, and the pentatricopeptide repeat-containing protein GUN1. The ultrafast bootstrap support of IQ-TREE2 is shown at each internal node; only values ≥ 70% are shown. Unit of branch length is the number of substitutions per site.

3. Conclusions

Our results demonstrate the power of long-read sequence data in elucidating key genome features in *C. goreaui*, at a comparable capacity to chromosome-level genome assemblies of other Symbiodiniaceae, including the resolution of duplicated genes, repetitive genomic elements, and TADs. These results support the expected high sequence and structural divergence of dinoflagellate genomes [9,10]. Comparative analysis of genes revealed clear evidence of lineage-specific innovation in *C. goreaui* and in dinoflagellates generally, implicating about one-half of *C. goreaui* genes; many (52.9%) *C. goreaui* genes remain dark, for which the encoded functions are unknown [24]. Our gene-by-gene phylogenetic analysis revealed the intricate evolutionary histories of genes in *C. goreaui* and dinoflagellates, with many too complex to be unambiguously interpreted.

Our results highlight how genetic transfer and gene duplication generated functional diversity and innovation in *C. goreaui*, and in combination with the conserved LTRs and DNA transposons, shaped the genome of this facultative symbiont [12]. The data generated from this study provide a useful reference for future studies of coral symbionts, and more broadly of dinoflagellates and microbial eukaryotes. The identified TAD regions, for instance, provide an excellent analysis platform to assess the presence of conserved gene-regulatory elements, e.g., promoters or enhancers of gene expression, as hypothesised in Lin et al. [22] to facilitate transcription of the unidirectional gene blocks. Our analytic workflow can be adapted and applied to study TADs in other assembled genomes of dinoflagellates.

4. Materials and Methods

4.1. Generation of Long-Read Genome and Transcriptome Data

Cladocopium goreaui SCF055-01 is a single-cell monoclonal culture first isolated from the coral *Acropora tenuis* at Magnetic Island, Queensland, Australia [51]. The cultures were maintained at the Australian Centre of Marine Science (AIMS) in Daigo's IMK medium at 26 °C, 90 µE/cm^2/s^{-1}. High molecular weight genomic DNA was extracted following the SDS method described in Wilson et al. [52]. The sample was sent to the Ramaciotti Centre for Genomics (University of New South Wales, Sydney) for sequencing using the PacBio, first using RS II, then the Sequel platform (Table S1). DNA fragments of lengths 10–20 Kb were selected for the preparation of sequencing libraries. In total, 6.2 million subreads were produced (50.2 Gbp).

Total RNA was extracted from cultured SCF055 cells in the exponential growth phase (~10^6 cells), combining the standard Trizol method with the Qiagen RNeasy protocol, following the method of Rosic and Hoegh-Guldberg [53]. Quality and quantity of RNA were assessed using a Bioanalyzer and Qubit. The RNA sample was sent to the sequencing facility at the University of Queensland's Institute for Molecular Bioscience for generation of Iso-Seq data using the PacBio Sequel II platform. The Iso-Seq library was prepared using the NEBNext Single Cell/Low Input cDNA Synthesis and Amplification Module and the SMRTbell Express Template Prep Kit 2.0, following standard protocol. Sequencing was conducted using half of a Sequel II SMRT cell. From these raw data, we generated 3,534,837 circular consensus sequencing (CCS) reads (7.3 Gb; average 36 passes) using CCS v4.2.0. The CCS reads were then fed into the Iso-Seq pipeline v3.3.0 pipeline for standard Iso-Seq processing, which includes key steps of read refinement, isoform clustering, and polishing, resulting in 55,505 high-quality, non-redundant, polished Iso-Seq transcripts (total 79 Mb; N50 length 1493 bases).

4.2. De Novo Genome Assembly Combining Short- and Long-Read Sequence Data

We combined the long-read sequence data with all short-read sequence data from Liu et al. [7] in a hybrid genome assembly using MaSuRCA v3.4.2 [54]. Because the SCF055 culture is xenic, we adapted the approach in Iha et al. [55] to identify and remove putative contaminant sequences from bacterial, archaeal, and viral sources. Bowtie2 [56] was first used to map the genome sequencing data (Illumina paired-end reads) using the *–very-fast* algorithm to the assembled genome scaffolds to obtain information of sequencing depth. BlobTools v1.1 [57] was then used to identify anomalies of GC content and sequencing depth among the scaffolds, and to assign a taxon to each scaffold (using the default *bestsum* algorithm) based on hits in a BLASTn ($E \leq 10^{-20}$) search against the NCBI nucleotide (nt) database (released 2020-01-08). We also mapped available transcriptome data onto the assembled genome to further assess gene structure to aid identification of intron-containing genes on the scaffolds as indication of eukaryote origin. To do this, we used our Iso-Seq transcripts (above) and the RNA-Seq data from Levin et al. [58] that we assembled using Trinity v2.9.1 [59] in de novo mode. Mapping was conducted using minimap2 v2.17-r975-dirty [60] (*–secondary=no -ax splice:hq -uf –splice-flank=no*), for which the code has been modified to recognise alternative splice-sites in dinoflagellate genes. Using the taxon assignment, genome coverage, and transcript support information, we identified and removed putative contaminant sequences from the genome assembly following a decision tree based on these results (Figure S8). Chloroplast and mitochondrial genome sequences were identified and subsequently removed from the final genome assembly, following the method of Stephens et al. [18].

4.3. Estimation of Genome Size Based on Sequencing Data

To estimate the genome size, we adapted the k-mer-based approach used by González-Pech et al. [10]. We first enumerated k-mers of size $k = 21$ from the sequencing reads using Jellyfish v2.3.0 [61] with the command *jellyfish histo –high=1,000,000*. The resulting histogram of k-mer count was used as input for GenomeScope2 [62] to estimate a haploid genome size. Genomes of *C. goreaui* (and other Symbiodiniaceae) are thought to be haploid [7], and we did not observe bimodal distribution of k-mer coverage expected in a diploid genome (Figure S1).

4.4. Annotation of Repeat Content

De novo repeat families were predicted from the genome assembly using RepeatModeler v1.0.11 (http://www.repeatmasker.org/RepeatModeler/ (accessed on 20 January 2021)). All repeats (including known repeats in RepeatMasker database release 26 October 2018) were identified and masked using RepeatMasker v4.0.7 (http://www.repeatmasker.org/ (accessed on 20 January 2021)); the masked sequences were used for ab initio gene prediction (below). The repeat landscape plot was generated with Perl script *createRepeatLandscape.pl* (available from RepeatMasker).

4.5. Ab Initio Prediction of Protein-Coding Genes

To predict protein-coding genes from the assembled genome sequences, we adopted the approach in Chen et al. [15], using the workflow tailored for dinoflagellate genomes (https://github.com/TimothyStephens/Dinoflagellate_Annotation_Workflow (accessed on 20 January 2021)), which was also applied in earlier studies of Symbiodiniaceae genomes [7,10]. Briefly, the transcriptome data (combining our 55,505 high-quality Iso-Seq transcripts and data from Levin et al. [58]; above) were mapped onto the assembled genome with Minimap2 [60]. All transcripts were combined into gene assemblies using PASA v2.3.3 [63], for which the code was modified to recognise alternative splice sites (available at https://github.com/chancx/dinoflag-alt-splice (accessed on 20 January 2021)). TransDecoder v5.2.0 [63] was used to predict open reading frames on the PASA-assembled transcripts; these represent the transcript-based predicted genes. The predicted proteins were searched (BLASTP, $E \leq 10^{-20}$, > 80% query cover) against a protein database consisting of RefSeq proteins (release 88) and predicted proteins of available Symbiodiniaceae genomes (Table S7). The gene models were checked for transposable elements using HHblits v2.0.16 [64] and TransposonPSI (http://transposonpsi.sourceforge.net/ (accessed on 20 January 2021)), searching against the JAMg transposon database (https://github.com/genomecuration/JAMg (accessed on 20 January 2021)); those containing these elements were removed from subsequent steps. After removal of redundant sequences based on similarity using CD-HIT v4.6.8 [65] (-c 0.75 -n 5), the final curated gene models were used to identify high-quality "golden genes" using the script *Prepare_golden_genes_for_predictors.pl* from the JAMg pipeline (http://jamg.sourceforge.net/ (accessed on 20 January 2021)), altered to recognise alternative splice sites.

We used four other programs for predicting protein-coding genes. The "golden genes" above were used as a training set for SNAP [66] and AUGUSTUS v3.3.1 [67] to predict genes from the repeat-masked genome; the code for AUGUSTUS was altered to recognise alternative splice sites of dinoflagellates (available at https://github.com/chancx/dinoflag-alt-splice (accessed on 20 January 2021)). We also used GeneMark-ES [68] and MAKER v2.31.10 [69], for which the code was modified to recognise alternative splice sites, in *protein2genome* mode guided by the SwissProt database (retrieved 27 June 2018) and other predicted proteins from Symbiodiniaceae (Table S7). Finally, gene predictions from all five methods, i.e., the ab initio predictions (from GeneMark-ES, AUGUSTUS, and SNAP), MAKER protein-based predictions, and PASA transcript-based predictions, were integrated using EvidenceModeler v1.1.1 [70] to yield the gene models (see Chen et al. [15] for detail). The gene models were further polished with PASA [63] for three rounds to incorporate isoforms and UTRs, yielding the final gene models.

4.6. Functional Annotation of C. goreaui Genes

For functional annotation, all predicted proteins were searched against all protein sequences on Uniport (release 2021_03). Only hits with $E \leq 10^{-5}$ were retained. Gene Ontology (GO) terms associated with top hits were first retrieved from the UniProt website using the *Retrieve/ID mapping* tool, then mapped to the corresponding queries.

4.7. Analysis of Duplicated Genes

To identify and classify duplicated genes in *C. goreaui*, we follow González-Pech et al. [10] to perform all-versus-all BLASTp ($E \leq 10^{-5}$) on corresponding proteins of all predicted genes. The top five hits (excluding the query itself) were used as input for MCScanX [71] in *duplicate_gene_classifier* mode to classify genes into five categories: singleton (single-copy genes), dispersed (paralogs away from each other; i.e., at least 20 genes apart), proximal (paralogs near each other), tandem (paralogs in tandem gene block), and segmental (duplicates of collinear blocks).

4.8. GO Enrichment Analysis

R package topGO [72] was used for enrichment analysis of GO terms. In total, 21,356 genes were annotated with one or more GO terms; these were used as the background set. Genes in tandem repeats and segmental duplication were used as the test set to search for enriched GO terms, in independent analyses. Fisher's exact test was applied to assess statistical significance, and instances with $p \leq 0.01$ were considered significant.

4.9. Analysis of Unidirectional Gene Blocks and TADs

For this part of analysis, we focused on five representative assembled genomes of dinoflagellates: *C. goreaui* from this study, *C. goreaui* from Chen et al. [15], *Fugacium kawagutii* [19], *Breviolum minutum* [15], and *Symbiodinium microadriaticum* [20]. Unidirectional gene blocks were identified based on a block of genes within which their orientations are the same. A putative TAD boundary is the region at which the orientations of two blocks converged. A putative TAD central region is the region at which the orientations diverged. We analysed putative TAD regions and unidirectional gene blocks based on the minimum number of genes within a block, N, at $N = 4, 6, 8,$ and 10.

To validate the putative TADs, we searched for GC dip in the TAD boundaries, following Nand et al. [20]. On each scaffold, for each sliding 4Kb-window, we calculated the localised G+C content. A putative region of GC dip is identified based on three criteria: (1) the G+C in a 4Kb region is lower than average GC content of the entire scaffold (i.e., the background); (2) the largest difference between the localised G+C and the background G+C is larger than 0.05%; and that (3) the implicated region is longer than 5000 bp.

4.10. Phylogenomic Analysis of C. goreaui Genes

To investigate the putative origins of *C. goreaui* genes, we compiled a comprehensive protein database (2,841,408 sequences) of 96 broadly sampled taxa from diverse lineages, encompassing eukaryotic, bacterial, and archaeal taxa, of which 30 are dinoflagellates (Table S5). For species where there were multiple datasets for the same isolate, the protein sequences were clustered at 90% sequence identity using CD-HIT-v4.8.1 [65] to reduce redundancy. Isoforms are reduced to retain one representative protein (longest) per gene.

Using all 2,841,408 protein sequences from the database (Table S5), homologous protein sets were inferred using OrthoFinder v2.3.8 [73] at the default setting. For this analysis, we restricted our analysis to 177,346 putative homologous sets of *C. goreaui* proteins (i.e., sets in which one or more *C. goreaui* sequence is represented). For homologous sets that contain only Dinophyceae and one other phylum (an exclusive gene-sharing partner), the putative gene-sharing partner was assessed based on the number of implicated homologous sets, requiring at least x number of genes in each set (for $x = 2, 20, 40$, and 60). For the other protein sets, multiple sequence alignment was performed using MAFFT-v7.453 [74] with parameters *–maxiterate 1000 –localpair*. Following the methods of Stephens et al. [24], ambiguous and non-phylogenetically informative sites in each alignment were trimmed using trimAl-v1.4.1 [75] in two steps: trimming directly with *-automated1*, then with *-resoverlap 0.5 -seqoverlap 50*. A maximum likelihood tree for each protein set was inferred from these trimmed alignments, using IQ-TREE2 [76], with an edge-unlinked partition model and 2000 ultrafast bootstraps. The initial step of IQ-TREE2 by default is to perform a composition chi-squared test for every sequence in an alignment, the sequence for which the character composition significantly deviates from the average composition in the alignment is removed. Alignments filtered this way were further removed if the target *C. goreaui* sequence was removed, and if the alignment contained no more than four sequences. In total, 5246 trees were used in subsequent analysis.

4.11. Inference of the Dinoflagellate Species Tree

To infer the dinoflagellate species tree, we first inferred homologous protein sets with OrthoFinder v2.3.8 for the 30 dinoflagellate taxa and *Perkinsus marinus* (outgroup) in Table S5. The 3411 strictly orthologous, single-copy protein sets (i.e., sets in which each dinoflagellate taxon is represented no more than once) were used for inferring the species tree. For each set, multiple sequence alignment was performed, and the alignment was trimmed, per our approach described above. A consensus maximum likelihood reference species tree was then inferred using IQ-TREE2, with an edge-unlinked partition model and 2000 ultrafast bootstraps.

4.12. Inference of C. goreaui Gene Origins

Putative origins of *C. goreaui* genes were determined based on the presence of strongly supported clades (determined at a bootstrap support threshold) that include *C. goreaui* (and/or other dinoflagellates) and one other taxon group (e.g., a phylum). We used PhySortR [31] to quickly sort through thousands of protein trees (i.e., we assume these as gene trees) for the specific target clades, independently at bootstrap thresholds of $\geq 90\%$ (more stringent, higher confidence), $\geq 70\%$, and $\geq 50\%$ (less stringent, lower confidence); default values were used for other parameters. Our 176-step tree-sorting strategy is detailed in Table S6. Briefly, we sorted the trees based on recovery of a strongly supported monophyletic clade containing both the subject group (dinoflagellates) and the target group in three stages: first (a) with target groups implicated in endosymbiosis in the evolution of plastid (e.g., Archaeplastida); then (b) with closely related target group expected under vertical inheritance; and finally (c) with other remotely related eukaryote or prokaryote target group as further indication of horizontal genetic transfer. At each stage, the subject group would proceed from the most inclusive (i.e., dinoflagellates plus other closely related taxa in the SAR supergroup, Cryptophyta, and Haptophyta), and progressively the most distant lineage was removed in each iterative sorting, leaving only dinoflagellates. For each subject group, the target group would proceed similarly from the most inclusive, e.g., in stage one, all three phyla of Archaeplastida, to subsequent individual phylum.

Supplementary Materials: The following supporting information can be downloaded at: https://www.mdpi.com/article/10.3390/microorganisms10081662/s1, Data S1: All sorted trees where Dinophyceae taxa form a strongly supported clade with one other phylum; Figure S1: Genome size estimation for *Cladocopium goreaui* using GenomeScope v2.0, based on frequency distribution of *k*-mers from short-read sequence data, shown for exact (top) and log-transformed (below) values; Figure S2: An example of G+C dip observed in the *C. goreaui* genome of a putative boundary of topologically associated domain (TAD), shown for 4000-bp sliding windows on scaffold scf7180000355754. The *x*-axis shows the centre position of each sliding window along the scaffold. Of the dashed lines, the red line indicates the mean %G+C of the scaffold (as background), the green lines signify the G+C dip region, and the black lines signify a putative TAD boundary; Figure S3: Species tree of 30 dinoflagellate taxa and *Perkinsus* as outgroup, inferred from 3411 strictly orthologous (single-copy) protein sets; Figure S4: Maximum likelihood tree showing gene expansion of a green algal derived protein family that contain a remote homolog in *Arabidopsis thaliana* with function implicated in cytokinesis and meiosis; Figure S5: Maximum likelihood tree of phosphatidylinositol 4-phosphate 5-kinase showing possible misidentification of the sequence from the dinoflagellate symbiont associated with the coral; Figure S6: Maximum likelihood tree of ubiquitin carboxyl-terminal hydrolase showing strong evidence of vertical inheritance; Figure S7: Domain configuration for a representative sequence from each of the three sub-clades in the tree of Figure 5, shown for the autophagy-related protein 18a, the transmembrane protein 43, and the pentatricopeptide repeat-containing protein GUN1; Figure S8: Decision tree for identification and removal of putative contaminant sequences; Table S1: PacBio long-read genome sequencing data from *C. goreaui* generated in this study; Table S2: Enriched GO terms for tandemly repeated genes, shown for Biological Process (BP), Cellular Component (CC), and Molecular Function (MF); Table S3: Enriched GO terms for genes in segmental duplicates, shown for Biological Process (BP), Cellular Component (CC), and Molecular Function (MF); Table S4: Enriched GO terms for genes disrupting unidirectional coding of gene blocks, shown for Biological Process (BP), Cellular Component (CC), and Molecular Function (MF); Table S5: Protein database for phylogenomic analysis; Table S6: Sorting order of phylogenetic trees and the associated classifications; Table S7: Protein sequences used to guide ab initio prediction of protein-coding genes.

Author Contributions: Conceptualization, Y.C., D.B. and C.X.C.; methodology, Y.C., S.S., K.E.D. and C.X.C.; formal analysis, Y.C. and S.S.; writing—original draft preparation, Y.C.; writing—review and editing, Y.C., M.J.H.v.O., D.B. and C.X.C.; supervision, C.X.C.; funding acquisition, C.X.C. and D.B. All authors have read and agreed to the published version of the manuscript.

Funding: This research was funded by Australian Research Council grants DP150101875 and DP190102474 (C.X.C. and D.B.), and the University of Queensland Research Training Program (Y.C. and S.S.). M.J.H.v.O. acknowledges the Australian Research Council Laureate Fellowship FL180100036.

Institutional Review Board Statement: Not applicable.

Informed Consent Statement: Not applicable.

Data Availability Statement: Genome and transcriptome long-read sequence data generated from this study are available at NCBI Sequence Read Archive (BioProject accession PRJEB55036). The assembled genome, predicted gene models and proteins, and the identified organellar genome sequences from this study are available at https://doi.org/10.48610/fba3259 (accessed on 30 June 2022).

Acknowledgments: This project is supported by high-performance computing facilities at the National Computational Infrastructure (NCI) National Facility systems through the NCI Merit Allocation Scheme (Project d85) awarded to C.X.C., the University of Queensland Research Computing Centre, and HPC from the Australian Centre for Ecogenomics at UQ. We are grateful for the technical assistance provided by Carlos Alvarez Roa and Lesa Peplow in the extractions of DNA and RNA from *C. goreaui* SCF055 at the Symbiont Culturing Facility at the Australian Institute of Marine Science in Townsville, Queensland.

Conflicts of Interest: The authors declare no conflict of interest. The funders had no role in the design of the study; in the collection, analyses, or interpretation of data; in the writing of the manuscript; or in the decision to publish the results.

References

1. LaJeunesse, T.C.; Parkinson, J.E.; Gabrielson, P.W.; Jeong, H.J.; Reimer, J.D.; Voolstra, C.R.; Santos, S.R. Systematic revision of Symbiodiniaceae highlights the antiquity and diversity of coral endosymbionts. *Curr. Biol.* **2018**, *28*, 2570–2580. [CrossRef] [PubMed]
2. Muscatine, L.; Falkowski, P.G.; Porter, J.W.; Dubinsky, Z.; Smith, D.C. Fate of photosynthetic fixed carbon in light-and shade-adapted colonies of the symbiotic coral *Stylophora pistillata*. *Proc. R. Soc. B Biol. Sci.* **1984**, *222*, 181–202.
3. Kopp, C.; Pernice, M.; Domart-Coulon, I.; Djediat, C.; Spangenberg, J.E.; Alexander, D.T.L.; Hignette, M.; Meziane, T.; Meibom, A.; Orphan, V.; et al. Highly dynamic cellular-level response of symbiotic coral to a sudden increase in environmental nitrogen. *mBio* **2013**, *4*, e00052-13. [CrossRef] [PubMed]
4. Hoegh-Guldberg, O. Climate change, coral bleaching and the future of the world's coral reefs. *Mar. Freshw. Res.* **1999**, *50*, 839–866. [CrossRef]
5. Lin, S.; Cheng, S.; Song, B.; Zhong, X.; Lin, X.; Li, W.; Li, L.; Zhang, Y.; Zhang, H.; Ji, Z. The *Symbiodinium kawagutii* genome illuminates dinoflagellate gene expression and coral symbiosis. *Science* **2015**, *350*, 691–694. [CrossRef] [PubMed]
6. Aranda, M.; Li, Y.; Liew, Y.J.; Baumgarten, S.; Simakov, O.; Wilson, M.C.; Piel, J.; Ashoor, H.; Bougouffa, S.; Bajic, V.B. Genomes of coral dinoflagellate symbionts highlight evolutionary adaptations conducive to a symbiotic lifestyle. *Sci. Rep.* **2016**, *6*, 39734. [CrossRef] [PubMed]
7. Liu, H.; Stephens, T.G.; González-Pech, R.A.; Beltran, V.H.; Lapeyre, B.; Bongaerts, P.; Cooke, I.; Aranda, M.; Bourne, D.G.; Forêt, S.; et al. *Symbiodinium* genomes reveal adaptive evolution of functions related to coral-dinoflagellate symbiosis. *Commun. Biol.* **2018**, *1*, 95.
8. Shoguchi, E.; Beedessee, G.; Tada, I.; Hisata, K.; Kawashima, T.; Takeuchi, T.; Arakaki, N.; Fujie, M.; Koyanagi, R.; Roy, M.C. Two divergent *Symbiodinium* genomes reveal conservation of a gene cluster for sunscreen biosynthesis and recently lost genes. *BMC Genom.* **2018**, *19*, 458. [CrossRef]
9. Dougan, K.E.; González-Pech, R.A.; Stephens, T.G.; Shah, S.; Chen, Y.; Ragan, M.A.; Bhattacharya, D.; Chan, C.X. Genome-powered classification of microbial eukaryotes: Focus on coral algal symbionts. *Trends Microbiol.* **2022**, *30*, 831–840. [CrossRef]
10. González-Pech, R.A.; Stephens, T.G.; Chen, Y.; Mohamed, A.R.; Cheng, Y.; Shah, S.; Dougan, K.E.; Fortuin, M.D.A.; Lagorce, R.; Burt, D.W.; et al. Comparison of 15 dinoflagellate genomes reveals extensive sequence and structural divergence in family Symbiodiniaceae and genus *Symbiodinium*. *BMC Biol.* **2021**, *19*, 73. [CrossRef]
11. Lo, R.; Dougan, K.E.; Chen, Y.; Shah, S.; Bhattacharya, D.; Chan, C.X. Alignment-free analysis of whole-genome sequences from Symbiodiniaceae reveals different phylogenetic signals in distinct regions. *Front. Plant Sci.* **2022**, *13*, 815714. [CrossRef] [PubMed]
12. González-Pech, R.A.; Bhattacharya, D.; Ragan, M.A.; Chan, C.X. Genome evolution of coral reef symbionts as intracellular residents. *Trends Ecol. Evol.* **2019**, *34*, 799–806. [CrossRef] [PubMed]
13. Bongaerts, P.; Carmichael, M.; Hay, K.B.; Tonk, L.; Frade, P.R.; Hoegh-Guldberg, O. Prevalent endosymbiont zonation shapes the depth distributions of scleractinian coral species. *R. Soc. Open Sci.* **2015**, *2*, 140297. [CrossRef]
14. LaJeunesse, T.C. "Species" radiations of symbiotic dinoflagellates in the Atlantic and Indo-Pacific since the Miocene-Pliocene transition. *Mol. Biol. Evol.* **2005**, *22*, 570–581. [CrossRef]
15. Chen, Y.; González-Pech, R.A.; Stephens, T.G.; Bhattacharya, D.; Chan, C.X. Evidence that inconsistent gene prediction can mislead analysis of dinoflagellate genomes. *J. Phycol.* **2020**, *56*, 6–10. [CrossRef] [PubMed]
16. Chan, C.X.; Soares, M.B.; Bonaldo, M.F.; Wisecaver, J.H.; Hackett, J.D.; Anderson, D.M.; Erdner, D.L.; Bhattacharya, D. Analysis of *Alexandrium tamarense* (Dinophyceae) genes reveals the complex evolutionary history of a microbial eukaryote *J. Phycol.* **2012**, *48*, 1130–1142. [CrossRef] [PubMed]
17. Manni, M.; Berkeley, M.R.; Seppey, M.; Simão, F.A.; Zdobnov, E.M. BUSCO update: Novel and streamlined workflows along with broader and deeper phylogenetic coverage for scoring of eukaryotic, prokaryotic, and viral genomes. *Mol. Biol. Evol.* **2021**, *38*, 4647–4654. [CrossRef]
18. Stephens, T.G.; González-Pech, R.A.; Cheng, Y.; Mohamed, A.R.; Burt, D.W.; Bhattacharya, D.; Ragan, M.A.; Chan, C.X. Genomes of the dinoflagellate *Polarella glacialis* encode tandemly repeated single-exon genes with adaptive functions. *BMC Biol.* **2020**, *18*, 56. [CrossRef]
19. Li, T.; Yu, L.; Song, B.; Song, Y.; Li, L.; Lin, X.; Lin, S. Genome improvement and core gene set refinement of *Fugacium kawagutii*. *Microorganisms* **2020**, *8*, 102. [CrossRef]
20. Nand, A.; Zhan, Y.; Salazar, O.R.; Aranda, M.; Voolstra, C.R.; Dekker, J. Genetic and spatial organization of the unusual chromosomes of the dinoflagellate *Symbiodinium microadriaticum*. *Nat. Genet.* **2021**, *53*, 618–629. [CrossRef]
21. Marinov, G.K.; Trevino, A.E.; Xiang, T.; Kundaje, A.; Grossman, A.R.; Greenleaf, W.J. Transcription-dependent domain-scale three-dimensional genome organization in the dinoflagellate *Breviolum minutum*. *Nat. Genet.* **2021**, *53*, 613–617. [CrossRef] [PubMed]
22. Lin, S.; Song, B.; Morse, D. Spatial organization of dinoflagellate genomes: Novel insights and remaining critical questions. *J. Phycol.* **2021**, *57*, 1674–1678. [CrossRef]

23. Price, D.C.; Bhattacharya, D. Robust Dinoflagellata phylogeny inferred from public transcriptome databases. *J. Phycol.* **2017**, *53*, 725–729. [CrossRef] [PubMed]
24. Stephens, T.G.; Ragan, M.A.; Bhattacharya, D.; Chan, C.X. Core genes in diverse dinoflagellate lineages include a wealth of conserved dark genes with unknown functions. *Sci. Rep.* **2018**, *8*, 17175. [CrossRef] [PubMed]
25. Burki, F.; Roger, A.J.; Brown, M.W.; Simpson, A.G.B. The new tree of eukaryotes. *Trends Ecol. Evol.* **2020**, *35*, 43–55. [CrossRef]
26. Hackett, J.D.; Yoon, H.S.; Li, S.; Reyes-Prieto, A.; Rümmele, S.E.; Bhattacharya, D. Phylogenomic analysis supports the monophyly of cryptophytes and haptophytes and the association of rhizaria with chromalveolates. *Mol. Biol. Evol.* **2007**, *24*, 1702–1713. [CrossRef]
27. Adl, S.M.; Bass, D.; Lane, C.E.; Lukeš, J.; Schoch, C.L.; Smirnov, A.; Agatha, S.; Berney, C.; Brown, M.W.; Burki, F.; et al. Revisions to the classification, nomenclature, and diversity of eukaryotes. *J. Eukaryot. Microbiol.* **2019**, *66*, 4–119. [CrossRef]
28. Ishida, K.-I.; Green, B.R. Second- and third-hand chloroplasts in dinoflagellates: Phylogeny of oxygen-evolving enhancer 1 (PsbO) protein reveals replacement of a nuclear-encoded plastid gene by that of a haptophyte tertiary endosymbiont. *Proc. Natl. Acad. Sci. USA* **2002**, *99*, 9294–9299. [CrossRef]
29. Chan, C.X.; Gross, J.; Yoon, H.S.; Bhattacharya, D. Plastid origin and evolution: New models provide insights into old problems. *Plant Physiol.* **2011**, *155*, 1552–1560. [CrossRef]
30. Yoon, H.S.; Hackett, J.D.; Van Dolah, F.M.; Nosenko, T.; Lidie, K.L.; Bhattacharya, D. Tertiary endosymbiosis driven genome evolution in dinoflagellate algae. *Mol. Biol. Evol.* **2005**, *22*, 1299–1308. [CrossRef]
31. Stephens, T.G.; Bhattacharya, D.; Ragan, M.A.; Chan, C.X. PhySortR: A fast, flexible tool for sorting phylogenetic trees in R. *PeerJ* **2016**, *4*, e2038. [CrossRef] [PubMed]
32. Gabrielsen, T.M.; Minge, M.A.; Espelund, M.; Tooming-Klunderud, A.; Patil, V.; Nederbragt, A.J.; Otis, C.; Turmel, M.; Shalchian-Tabrizi, K.; Lemieux, C. Genome evolution of a tertiary dinoflagellate plastid. *PLoS ONE* **2011**, *6*, e19132. [CrossRef] [PubMed]
33. Janouškovec, J.; Horák, A.; Oborník, M.; Lukeš, J.; Keeling, P.J. A common red algal origin of the apicomplexan, dinoflagellate, and heterokont plastids. *Proc. Natl. Acad. Sci. USA* **2010**, *107*, 10949–10954. [CrossRef] [PubMed]
34. Moustafa, A.; Beszteri, B.; Maier, U.G.; Bowler, C.; Valentin, K.; Bhattacharya, D. Genomic footprints of a cryptic plastid endosymbiosis in diatoms. *Science* **2009**, *324*, 1724–1726. [CrossRef]
35. Chan, C.X.; Reyes-Prieto, A.; Bhattacharya, D. Red and green algal origin of diatom membrane transporters: Insights into environmental adaptation and cell evolution. *PLoS ONE* **2011**, *6*, e29138. [CrossRef]
36. Nevo, Z.; Sharon, N. The cell wall of *Peridinium westii*, a non cellulosic glucan. *Biochim. Biophys. Acta* **1969**, *173*, 161–175. [CrossRef]
37. Salmeán, A.A.; Duffieux, D.; Harholt, J.; Qin, F.; Michel, G.; Czjzek, M.; Willats, W.G.T.; Hervé, C. Insoluble (1→3), (1→4)-β-D-glucan is a component of cell walls in brown algae (Phaeophyceae) and is masked by alginates in tissues. *Sci. Rep.* **2017**, *7*, 2880. [CrossRef]
38. Footitt, S.; Douterelo-Soler, I.; Clay, H.; Finch-Savage, W.E. Dormancy cycling in *Arabidopsis* seeds is controlled by seasonally distinct hormone-signaling pathways. *Proc. Natl. Acad. Sci. USA* **2011**, *108*, 20236–20241. [CrossRef]
39. Krochko, J.E.; Abrams, G.D.; Loewen, M.K.; Abrams, S.R.; Cutler, A.J. (+)-Abscisic acid 8′-hydroxylase is a cytochrome P450 monooxygenase. *Plant Physiol.* **1998**, *118*, 849–860. [CrossRef]
40. Deng, Y.; Hu, Z.; Shang, L.; Peng, Q.; Tang, Y.Z. Transcriptomic analyses of *Scrippsiella trochoidea* reveals processes regulating encystment and dormancy in the life cycle of a dinoflagellate, with a particular attention to the role of abscisic acid. *Front. Microbiol.* **2017**, *8*, 2450. [CrossRef]
41. Rosic, N.N.; Pernice, M.; Dunn, S.; Dove, S.; Hoegh-Guldberg, O. Differential regulation by heat stress of novel cytochrome P450 genes from the dinoflagellate symbionts of reef-building corals. *Appl. Environ. Microbiol.* **2010**, *76*, 2823–2829. [CrossRef] [PubMed]
42. Melis, A.; Chen, H.-C. Chloroplast sulfate transport in green algae–genes, proteins and effects. *Photosynth. Res.* **2005**, *86*, 299–307. [CrossRef] [PubMed]
43. Depuydt, T.; Vandepoele, K. Multi-omics network-based functional annotation of unknown *Arabidopsis* genes. *Plant J.* **2021**, *108*, 1193–1212. [CrossRef] [PubMed]
44. Kaundal, R.; Saini, R.; Zhao, P.X. Combining machine learning and homology-based approaches to accurately predict subcellular localization in *Arabidopsis*. *Plant Physiol.* **2010**, *154*, 36–54. [CrossRef]
45. Sarai, C.; Tanifuji, G.; Nakayama, T.; Kamikawa, R.; Takahashi, K.; Yazaki, E.; Matsuo, E.; Miyashita, H.; Ishida, K.-I.; Iwataki, M.; et al. Dinoflagellates with relic endosymbiont nuclei as models for elucidating organellogenesis. *Proc. Natl. Acad. Sci. USA* **2020**, *117*, 5364–5375. [CrossRef]
46. Yoon, H.S.; Hackett, J.D.; Bhattacharya, D. A single origin of the peridinin- and fucoxanthin-containing plastids in dinoflagellates through tertiary endosymbiosis. *Proc. Natl. Acad. Sci. USA* **2002**, *99*, 11724–11729. [CrossRef]
47. Morozov, A.A.; Galachyants, Y.P. Diatom genes originating from red and green algae: Implications for the secondary endosymbiosis models. *Mar. Genom.* **2019**, *45*, 72–78. [CrossRef]
48. Archibald, J.M.; Keeling, P.J. Recycled plastids: A 'green movement' in eukaryotic evolution. *Trends Genet.* **2002**, *18*, 577–584. [CrossRef]
49. Kamikawa, R.; Tanifuji, G.; Kawachi, M.; Miyashita, H.; Hashimoto, T.; Inagaki, Y. Plastid genome-based phylogeny pinpointed the origin of the green-colored plastid in the dinoflagellate *Lepidodinium chlorophorum*. *Genome Biol. Evol.* **2015**, *7*, 1133–1140. [CrossRef]

50. Rowan, R.; Whitney, S.M.; Fowler, A.; Yellowlees, D. Rubisco in marine symbiotic dinoflagellates: Form II enzymes in eukaryotic oxygenic phototrophs encoded by a nuclear multigene family. *Plant Cell* **1996**, *8*, 539–553.
51. Howells, E.; Beltran, V.; Larsen, N.; Bay, L.; Willis, B.; Van Oppen, M. Coral thermal tolerance shaped by local adaptation of photosymbionts. *Nat. Clim. Chang.* **2012**, *2*, 116–120. [CrossRef]
52. Wilson, K.; Li, Y.; Whan, V.; Lehnert, S.; Byrne, K.; Moore, S.; Pongsomboon, S.; Tassanakajon, A.; Rosenberg, G.; Ballment, E.; et al. Genetic mapping of the black tiger shrimp *Penaeus monodon* with amplified fragment length polymorphism. *Aquaculture* **2002**, *204*, 297–309. [CrossRef]
53. Rosic, N.N.; Hoegh-Guldberg, O. A method for extracting a high-quality RNA from *Symbiodinium* sp. *J. Appl. Phycol.* **2010**, *22*, 139–146. [CrossRef]
54. Zimin, A.V.; Puiu, D.; Luo, M.-C.; Zhu, T.; Koren, S.; Marçais, G.; Yorke, J.A.; Dvořák, J.; Salzberg, S.L. Hybrid assembly of the large and highly repetitive genome of *Aegilops tauschii*, a progenitor of bread wheat, with the MaSuRCA mega-reads algorithm. *Genome Res.* **2017**, *27*, 787–792. [CrossRef] [PubMed]
55. Iha, C.; Dougan, K.E.; Varela, J.A.; Avila, V.; Jackson, C.J.; Bogaert, K.A.; Chen, Y.; Judd, L.M.; Wick, R.; Holt, K.E. Genomic adaptations to an endolithic lifestyle in the coral-associated alga *Ostreobium*. *Curr. Biol.* **2021**, *31*, 1393–1402. [CrossRef] [PubMed]
56. Langmead, B.; Salzberg, S.L. Fast gapped-read alignment with Bowtie 2. *Nat. Methods* **2012**, *9*, 357–359. [CrossRef]
57. Laetsch, D.; Blaxter, M. BlobTools: Interrogation of genome assemblies. *F1000Research* **2017**, *6*, 1287. [CrossRef]
58. Levin, R.A.; Beltran, V.H.; Hill, R.; Kjelleberg, S.; McDougald, D.; Steinberg, P.D.; van Oppen, M.J.H. Sex, scavengers, and chaperones: Transcriptome secrets of divergent *Symbiodinium* thermal tolerances. *Mol. Biol. Evol.* **2016**, *33*, 2201–2215. [CrossRef]
59. Grabherr, M.G.; Haas, B.J.; Yassour, M.; Levin, J.Z.; Thompson, D.A.; Amit, I.; Adiconis, X.; Fan, L.; Raychowdhury, R.; Zeng, Q. Trinity: Reconstructing a full-length transcriptome without a genome from RNA-Seq data. *Nat. Biotechnol.* **2011**, *29*, 644. [CrossRef]
60. Li, H. Minimap2: Pairwise alignment for nucleotide sequences. *Bioinformatics* **2018**, *34*, 3094–3100. [CrossRef]
61. Marçais, G.; Kingsford, C. A fast, lock-free approach for efficient parallel counting of occurrences of k-mers. *Bioinformatics* **2011**, *27*, 764–770. [CrossRef] [PubMed]
62. Ranallo-Benavidez, T.R.; Jaron, K.S.; Schatz, M.C. GenomeScope 2.0 and Smudgeplot for reference-free profiling of polyploid genomes. *Nat. Commun.* **2020**, *11*, 1432. [CrossRef] [PubMed]
63. Haas, B.J.; Delcher, A.L.; Mount, S.M.; Wortman, J.R.; Smith, R.K., Jr.; Hannick, L.I.; Maiti, R.; Ronning, C.M.; Rusch, D.B.; Town, C.D. Improving the *Arabidopsis* genome annotation using maximal transcript alignment assemblies. *Nucleic Acids Res.* **2003**, *31*, 5654–5666. [CrossRef] [PubMed]
64. Remmert, M.; Biegert, A.; Hauser, A.; Söding, J. HHblits: Lightning-fast iterative protein sequence searching by HMM-HMM alignment. *Nat. Methods* **2012**, *9*, 173–175. [CrossRef] [PubMed]
65. Li, W.; Godzik, A. Cd-hit: A fast program for clustering and comparing large sets of protein or nucleotide sequences. *Bioinformatics* **2006**, *22*, 1658–1659. [CrossRef] [PubMed]
66. Korf, I. Gene finding in novel genomes. *BMC Bioinform.* **2004**, *5*, 59. [CrossRef]
67. Stanke, M.; Keller, O.; Gunduz, I.; Hayes, A.; Waack, S.; Morgenstern, B. AUGUSTUS: Ab *initio* prediction of alternative transcripts. *Nucleic Acids Res.* **2006**, *34*, W435–W439. [CrossRef]
68. Lomsadze, A.; Ter-Hovhannisyan, V.; Chernoff, Y.O.; Borodovsky, M. Gene identification in novel eukaryotic genomes by self-training algorithm. *Nucleic Acids Res.* **2005**, *33*, 6494–6506. [CrossRef]
69. Holt, C.; Yandell, M. MAKER2: An annotation pipeline and genome-database management tool for second-generation genome projects. *BMC Bioinform.* **2011**, *12*, 491. [CrossRef]
70. Haas, B.J.; Salzberg, S.L.; Zhu, W.; Pertea, M.; Allen, J.E.; Orvis, J.; White, O.; Buell, C.R.; Wortman, J.R. Automated eukaryotic gene structure annotation using EVidenceModeler and the Program to Assemble Spliced Alignments. *Genome Biol.* **2008**, *9*, R7. [CrossRef]
71. Wang, Y.; Li, J.; Paterson, A.H. MCScanX-transposed: Detecting transposed gene duplications based on multiple colinearity scans. *Bioinformatics* **2013**, *29*, 1458–1460. [CrossRef] [PubMed]
72. Alexa, A.; Rahnenführer, J. topGO: Enrichment analysis for Gene Ontology. R package version 2.48.0. 2022. Available online: https://bioconductor.org/packages/release/bioc/html/topGO.html (accessed on 30 June 2022).
73. Emms, D.M.; Kelly, S. OrthoFinder: Phylogenetic orthology inference for comparative genomics. *Genome Biol.* **2019**, *20*, 238. [CrossRef] [PubMed]
74. Katoh, K.; Standley, D.M. MAFFT multiple sequence alignment software version 7: Improvements in performance and usability. *Mol. Biol. Evol.* **2013**, *30*, 772–780. [CrossRef] [PubMed]
75. Capella-Gutiérrez, S.; Silla-Martínez, J.M.; Gabaldón, T. trimAl: A tool for automated alignment trimming in large-scale phylogenetic analyses. *Bioinformatics* **2009**, *25*, 1972–1973. [CrossRef] [PubMed]
76. Minh, B.Q.; Schmidt, H.A.; Chernomor, O.; Schrempf, D.; Woodhams, M.D.; von Haeseler, A.; Lanfear, R. IQ-TREE 2: New models and efficient methods for phylogenetic inference in the genomic era. *Mol. Biol. Evol.* **2020**, *37*, 1530–1534. [CrossRef]

microorganisms

MDPI

Article

Metagenomic Analysis of the Species Composition and Seasonal Distribution of Marine Dinoflagellate Communities in Four Korean Coastal Regions

Jinik Hwang [1], Hee Woong Kang [1], Seung Joo Moon [2], Jun-Ho Hyung [2], Eun Sun Lee [2] and Jaeyeon Park [2,*]

[1] West Sea Fisheries Research Institute, National Institute of Fisheries Science, Incheon 22383, Korea; jinike12@korea.kr (J.H.); hwgang@korea.kr (H.W.K.)
[2] Environment and Resource Convergence Center, Advanced Institute of Convergence Technology, Suwon 16229, Korea; sjmoon04@snu.ac.kr (S.J.M.); hjh1120@snu.ac.kr (J.-H.H.); eunsun742@snu.ac.kr (E.S.L.)
* Correspondence: bada0@snu.ac.kr; Tel.: +82-31-888-9042; Fax: +82-31-888-9040

Abstract: Biomonitoring of dinoflagellate communities in marine ecosystems is essential for efficient water quality management and limiting ecosystem disturbances. Current identification and monitoring of toxic dinoflagellates, which cause harmful algal blooms, primarily involves light or scanning electron microscopy; however, these techniques are limited in their ability to monitor dinoflagellates and plankton, leaving an incomplete analysis. In this study, we analyzed the species composition and seasonal distribution of the dinoflagellate communities in four Korean coastal regions using 18S rRNA amplicon sequencing. The results showed significantly high diversity in the dinoflagellate communities in all regions and seasons. Furthermore, we found seasonally dominant species and causative species of harmful algal blooms (*Cochlodinium* sp., *Alexandrium* sp., *Dinophysis* sp., and *Gymnodinium* sp.). Moreover, dominant species were classified by region and season according to the difference in geographical and environmental parameters. The molecular analysis of the dinoflagellate community based on metagenomics revealed more diverse species compositions that could not be identified by microscopy and revealed potentially harmful or recently introduced dinoflagellate species. In conclusion, metagenomic analysis of dinoflagellate communities was more precise and obtained results faster than microscopic analysis, and could improve the existing monitoring techniques for community analysis.

Keywords: dinoflagellates; metagenomics; next-generation sequencing; monitoring

Citation: Hwang, J.; Kang, H.W.; Moon, S.J.; Hyung, J.-H.; Lee, E.S.; Park, J. Metagenomic Analysis of the Species Composition and Seasonal Distribution of Marine Dinoflagellate Communities in Four Korean Coastal Regions. *Microorganisms* **2022**, *10*, 1459. https://doi.org/10.3390/microorganisms10071459

Academic Editor: Simon K. Davy

Received: 29 April 2022
Accepted: 18 July 2022
Published: 19 July 2022

Publisher's Note: MDPI stays neutral with regard to jurisdictional claims in published maps and institutional affiliations.

1. Introduction

Marine dinoflagellates are ubiquitous and play diverse roles in marine ecosystems [1,2]. Some dinoflagellate species can grow out of control due to various environmental factors, such as excessive inorganic nutrients (nitrogen (N) and phosphorus (P)) introduced from the land, forming a bloom [3,4]. Blooms from dinoflagellates have detrimental effects on a variety of aquatic animals, including fish and aquatic mammals, and can even be harmful to humans through toxin production [5,6]. Therefore, continuous monitoring of dinoflagellate communities is essential, as they can affect the diversity of surrounding aquatic life and cause ecosystem disturbance.

To date, monitoring of dinoflagellates in the aquatic environment has generally involved morphological identification using light microscopy observations. Recent advances in microscopy, including scanning electron microscopy (SEM), have enabled more precise identification [7,8]. However, the morphological classification of plankton via microscopy is still challenging, as plankton are difficult to observe with SEM due to the lack of an outer shell in dinoflagellates or the extremely small size of plankton. Recently, many types of species identification technology to distinguish dinoflagellates and molecular technology targeting species-specific genes have been developed [9,10]. In particular, next-generation sequencing (NGS) has greatly expanded our understanding of the diversity and function of dinoflagellates in the aquatic environment. This technique allows for rapid, high-resolution analysis of microbial and dinoflagellate communities [11,12]. In addition, it is possible to accurately identify nano- and pico-sized plankton, which are difficult to distinguish with a conventional microscope, facilitating the identification of various plankton that have been overlooked because they do not appear or are difficult to distinguish in local environmental conditions [13]. Although the QIIME or USEARCH pipeline has been widely used to analyze 16S rRNA gene sequencing reads from microbial communities [14–16], many metagenomics studies examining the profile of marine dinoflagellates have been carried out using the CLC Genomics Workbench [17–20]. In this study, we analyzed taxonomic profiling and seasonal distribution of the dinoflagellate communities in four Korean coastal regions based on the reading of 18S rRNA sequences using the CLC Workbench. To verify the results calculated using the CLC tool, those results were compared with abundance measured by direct counting of cells using microscopy.

Outbreaks of harmful dinoflagellates have traditionally occurred in tropical or temperate regions which have the potential for enhancing the growth rate of phytoplankton cells under the appropriate environmental conditions. Jeju Island, located along the southern coast of Korea, is a temperate region, and the occurrence of benthic dinoflagellates producing phytotoxins has been frequently reported in Jeju [21]. Understanding the spatial and seasonal dynamics of the toxic dinoflagellates in this region is essential, and many researchers have continuously monitored the cell abundance around Jeju Island using microscopic identification [22–24]. In this study, we investigated the spatial and seasonal variation of dinoflagellate communities in four different sites in Korean coastal waters, including Jeju Island, using NGS-based (18S rRNA amplicon) metagenomics. For precise bioinformatics analyses, we established a reference database of dinoflagellates and analyzed the precision of NGS compared to conventional microscopic observation. Thus, the reference data for the dinoflagellate community classified based on the NGS findings in this study will provide a better understanding of the occurrence of toxic dinoflagellates in Korea.

2. Materials and Methods

2.1. Study Areas and Seawater Sample Collecting

Seawater samples for metagenomic analysis were collected from four coastal waters (Gunsan, Pohang, Tongyeong, and Seongsan) in March, June, September, and December 2019. The four selected sampling sites have different geological and environmental characteristics, representing the eastern coast (Pohang), southern coast (Tongyeong), western coast (Gunsan), and Jeju island (Seongsan), and all four locations are near a port with considerable human activity (Figure 1a). To remove large zooplankton and foreign substances in the sample, surface seawater at each region was sieved using meshes with pore sizes of 80 μm. Four liters of seawater samples for metagenomics analysis were filtered through a polycarbonate filter membrane (0.8 μm Millipore; MilliporeSigma, Burlington, MA, USA) to obtain environmental DNA samples, then transferred to the laboratory on dry ice. For microscopic analysis, 500 mL of seawater samples was fixed with Lugol's solution, and phytoplankton cells were identified to at least the genus level using an optical microscope (Axioskop; Zeiss, Oberkochen, Germany). The dinoflagellate cells were counted directly using a Sedgwick-Rafter counting chamber by light microscopy (BX53; Olympus, Tokyo, Japan). Environmental data, such as water temperature, pH, dissolved oxygen, and conductivity, were measured at each location using a YSI 566 Multi Probe System (YSI Inc., Yellow Springs, OH, USA).

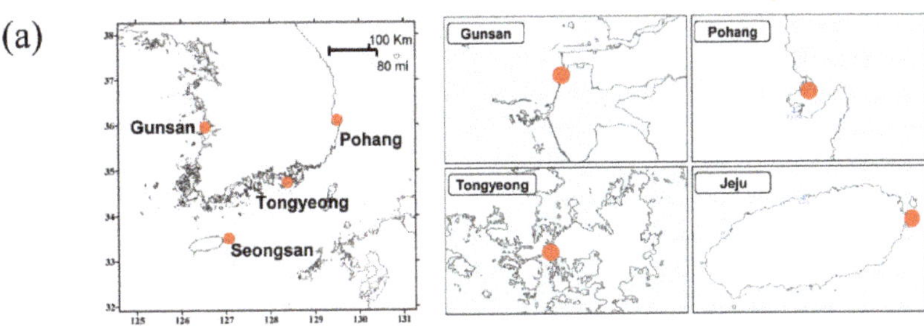

(a)

(b)

	Season	Temperature (°C)	Salinity (‰)	pH	DO (mg/L)
Gunsan	March	9.8	31.4	8.22	8.33
	June	18.5	30.1	8.12	8.24
	September	26.3	31.2	8.36	8.36
	December	6.3	24.9	8.41	8.46
Pohang	March	10.8	33.2	8.22	8.21
	June	16.4	33.9	8.22	9.11
	September	20.1	31.4	8.16	9.21
	December	10.3	34.7	8.24	9.04
Tongyeong	March	9.4	33.9	8.28	9.31
	June	21.1	33.2	8.27	9.27
	September	24.5	32.2	8.17	9.09
	December	12.1	32.2	8.23	9.17
Seongsan	March	14.8	34.3	7.89	8.43
	June	18.5	33.9	7.98	8.32
	September	24.3	27.9	7.82	9.04
	December	13.7	34.8	7.96	9.07

Figure 1. Location of sample sites and environmental indices at these sites (**a**) in four regions of Korean coastal waters (**b**).

2.2. DNA Extraction, Library Preparation, and NGS

DNA was extracted from the filtered membranes containing dinoflagellates and microbial cells using a DNeasy PowerSoil Kit (Qiagen, Hilden, Germany) following the manufacturer's instructions. The amount of double-stranded DNA and the purity in the extracted DNA samples was measured by PicoGreen (Promega, Madison, WI, USA) using VICTOR Nivo (PerkinElmer, Waltham, MA, USA). Per the Illumina 16S Metagenomic Sequencing Library protocols, the V3-V4 region of 18S ribosomal DNA (rDNA) gene in each sample was amplified by PCR using the following primers: 18S amplicon PCR forward primer, 5′–TCGTCGGCAGCGTCAGATGTGTATAAGAGACAGCCAGCASCYGC GGTAATTCC-3′, reverse primer, 5′–GTCTCGTGGGCTCGGAGATGTGTATAAG -AGACAGACTTTCG-TTCTTGATYRA-3′ [25]. A subsequent amplification step with limited-cycle reaction was performed to add multiplexing indices and Illumina sequencing adapters. The PCR products were pooled, cleaned, and normalized using the PicoGreen, and the size of libraries was measured using a TapeStation DNA screen tape D1000 (Agilent Technologies, Santa

Clara, CA, USA). Sequence libraries in the sample were verified using the MiSeq™ platform (Illumina, San Diego, CA, USA).

2.3. Customized Dinoflagellate Reference Databases for CLC Workflows

For the DNA reference databases of dinoflagellates, a list of 1555 species of dinoflagellates named in a previous study [26] was prepared in the form of Excel data, and the reference database deposited in the NCBI was additionally downloaded. A total of approximately 5000 dinoflagellate reference databases were retrieved. The files were imported into CLC and customized for use as databases specified for analyzing dinoflagellate species. The analysis program used in this study was CLC Genomics Workbench 21.0.4 with CLC Microbial Genomics Module 21.0 (CLC Bio, Qiagen Company, Aarhus, Denmark) and was used for future species identification (Figure S1).

2.4. Data Quality Control and Taxonomic Profiling

Data quality control and taxonomic profiling were performed using the CLC Microbial Genomics Module (MGM). First, Reads were trimmed using the Trim Reads tool. The percentage of trimmed from approximately 300,000 reads per sample was 71% (n = 16). We trimmed the 5′ and 3′ terminal nucleotides of the reads, and discarded unqualified reads showing that the quality limit was less than 0.001 or ambiguous nucleotides were more than two. The average length of reads after trimming was between 217–234 bp. Samples with less than 100 reads (minimum percent from the median = 50.0) were removed. Second, the remaining qualified reads were used for operational taxonomic unit (OTU) clustering based on SILVA 18s v132 Database including 1555 dinoflagellates at a 97% sequence similarity. The detected chimeric sequences and singletons (Chimera crossover cost = 3, K-mer size = 6) were discarded. A phylogenetic tree using the neighbor-joining method with 100 replicates was constructed based on the aligned OTU sequences by the MUSCLE tool v3.8.425. The phylogeny was applied for alpha and beta diversity measures. The beta diversity was measured using the Euclidean distance, and principal coordinate analysis (PCoA) based on a Bray–Curtis dissimilarity matrix was performed to illustrate a hierarchical clustering heat map showing the correlation between the examined samples.

3. Results

3.1. Environmental Characteristics of Sampling Sites

The four selected sampling sites had different geological and environmental characteristics. All the regions showed four distinct seasons; however, there was a regional difference in water temperature. The month of March showed the lowest water temperature (6.4–14.3 °C) throughout the region, and September (20.1–26.3 °C) showed the highest water temperature. On average, the water temperature at Jeju Island (Seongsan) was higher than that of the land. The salinity did not show a significant difference by region (31.4–33.7‰), and the pH and dissolved oxygen amount also did not show significant regional changes (Figure 1b).

3.2. Metagenome Comparisons

A pipeline for metagenomics analysis of environmental DNA samples was developed to address the identification of dinoflagellates species. On average, over 300,000 reads were acquired from each region using the MiSeq™ platform (Illumina, San Diego, CA, USA), with a read length of 301 bp. After quality trimming and filtering of reads, 70.3% of the raw reads remained (Figure 2a), with an overall higher G+C content for reads obtained from the library.

(a)

Location	Month	Number of reads	Avg. reads length	Number of reads after trim	Percentage Trimmed (%)
Gunsan	Mar.	369,228	301	286,332	77.55
	Jun.	372,760	301	284,221	76.25
	Sep.	356,460	301	264,331	74.15
	Dec.	392,078	301	302,114	77.05
Pohang	Mar.	386,724	301	284,332	73.52
	Jun.	364,238	301	274,665	75.41
	Sep.	335,390	301	266,443	79.44
	Dec.	372,264	301	246,554	66.23
Tongyeong	Mar.	346,221	301	264,558	76.41
	Jun.	364,522	301	275,852	75.67
	Sep.	366,955	301	286,441	78.06
	Dec.	299,622	301	224,332	74.87
Seongsan	Mar.	321,552	301	210,441	65.45
	Jun.	366,541	301	222,224	60.63
	Sep.	325,226	301	240,225	73.86
	Dec.	263,325	301	200,224	76.04

(b)

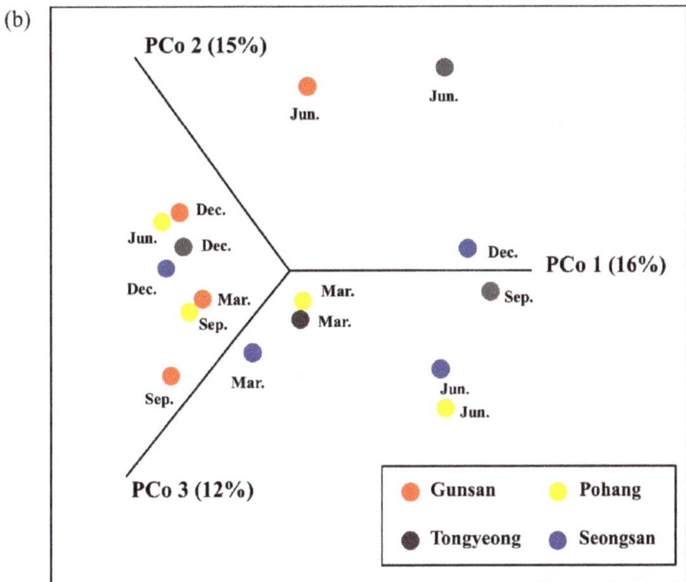

Figure 2. Comparison of metagenome libraries. Next-generation sequencing metadata including number of reads and trimmed reads (**a**), β-diversity (principal coordinate analysis (PCoA), dinoflagellate genotype composition (proportions) was measured by Bray–Curtis distances (**b**).

The nucleotide sequence similarity of the dinoflagellate genes was expressed by region using PCoA to illustrate the overall regional similarity according to the season. The December samples for Gunsan, Tongyeong, and Seongsan showed similarities, and the March samples of Pohang, Tongyeong, and Seongsan were also similar. The June and September samples of Tongyeong, in which a single species bloomed and became dominant, showed no similarity with the other samples. Furthermore, low similarity was found at Gunsan in June compared with the other samples (Figure 2b).

3.3. Metagenomic Analysis of the Dinoflagellate Species Composition

To identify marine dinoflagellates, we used the CLC genomics workbench program (CLC Microbial Genomics Module) on the assembled read sequences, followed by BLAST searches on the NCBI database and the newly created database of 1555 dinoflagellate species. Following the metagenomic analysis, 64 species of dinoflagellate were found in all regions on average. The top 10 dinoflagellates were selected based on the analyzed reads (Table 1).

Table 1. Seasonal variations and distribution of dinoflagellates in four coastal waters (Gunsan, Pohang, Tongyeong, and Seongsan) by metagenomic analysis. Total dinoflagellate reads and unidentified reads (**a**), and proportion(%) of the 10 most common dinoflagellate species (**b**).

(a)

Location	March		June		September		December	
	Dinoflagellate Reads	Unidentified	Dinoflagellate Reads	Unidentified	Dinoflagellate Reads	Unidentified	Dinoflagellate Reads	Unidentified
Gunsan	47,588	16,263	36,787	3084	39,780	14,265	87,273	11,245
Pohang	34,641	16,445	76,439	12,304	7222	4621	75,121	14,332
Tongyeong	12,397	3606	22,549	2855	69,468	1425	69,819	14,224
Seongsan	8924	2060	60,769	8994	30,197	4962	42,927	8644

(b)

Location	March		June		September		December	
	Proportion (%)	Species	Proportion (%)	Species	Proportion (%)	Species	Proportion (%)	Species
Gunsan	32.1	*Karlodinium veneficum*	45.2	*Gonyaulax* sp.	24.1	*Karlodinium* sp.	34.2	*Gyrodinium* sp.
	24.2	*Gyrodinium* sp.	15.3	*Symbiodinium* sp.	17.6	*Akashiwo* sp.	24.8	*Amphidiniella* sp.
	7.9	*Gymnodinium* sp.	7.9	*Karlodinium veneficum*	5.5	*Sinophysis* sp.	10.5	*Ceratium* sp.
	0.4	*Noctiluca scintillans*	6.3	*Ceratium* sp.	4.2	*Peridinium* sp.	4.5	*Heterocapsa triquetra*
	0.3	*Symbiodinium* sp.	3.1	*Pelagodinium* sp.	2.9	*Scrippsiella trochoidea*	2.5	*Karlodinium* sp.
	0.3	*Protoperidinium* sp.	2.4	*Dissodinium pseudolunula*	2.5	*Katodinium* sp.	2.4	*Peridinium* sp.
	0.3	*Pelagodinium* sp.	2.4	*Alexandrium* sp.	1.7	*Pelagodinium* sp.	2.1	*Noctiluca scintillans*
	0.2	*Scrippsiella* sp.	1.7	*Gyrodinium* sp.	1.3	*Gyrodinium* sp.	1.3	*Katodinium* sp.
	0.2	*Dinophysis* sp.	1.2	*Azadinium* sp.	0.8	*Cochlodinium* sp.	0.9	*Akashiwo* sp.
	0.2	*Ceratium* sp.	1.1	*Amphidiniopsis* sp.	0.5	*Ceratium* sp.	0.7	*Gonyaulax* sp.
Pohang	19.0	*Katodinium* sp.	52.1	*Gyrodinium* sp.	16.0	*Karlodinium* sp.	27.2	*Gyrodinium* sp.
	10.1	*Gyrodinium* sp.	4.9	*Heterocapsa triquetra*	6.8	*Sinophysis* sp.	16.2	*Bysmatrum arenicola*
	6.9	*Gymnodinium* sp.	4.0	*Ceratium* sp.	4.1	*Akashiwo* sp.	7.7	*Karlodinium veneficum*
	5.4	*Azadinium* sp.	3.7	*Karlodinium* sp.	1.5	*Paragymnodinium* sp.	7.4	*Akashiwo* sp.
	3.3	*Dinophysis* sp.	2.8	*Heterocapsa circularisquama*	1.5	*Peridinium* sp.	7.3	*Ceratium* sp.
	1.5	*Pelagodinium* sp.	2.1	*Gonyaulax* sp.	1.0	*Amphidiniella* sp.	4.6	*Cochlodinium* sp.
	1.4	*Ceratium* sp.	1.9	*Pelagodinium* sp.	1.0	*Ceratium* sp.	2.9	*Azadinium* sp.
	0.9	*Gonyaulax spinifera*	1.1	*Prorocentrum* sp.	0.5	*Bysmatrum arenicola*	1.5	*Katodinium* sp.
	0.9	*Gonyaulax* sp.	0.9	*Peridinium* sp.	0.4	*Pelagodinium* sp.	0.6	*Alexandrium* sp.
	0.8	*Erythropsidinium* sp.	0.6	*Cochlodinium* sp.	0.4	*Scrippsiella trochoidea*	0.6	*Peridinium* sp.
Tong-yeong	23.9	*Gyrodinium* sp.	50.2	*Prorocentrum* sp.	77.3	*Cochlodinium* sp.	62.3	*Gyrodinium* sp.
	23.6	*Gymnodinium* sp.	5.9	*Gyrodinium* sp.	7.9	*Gyrodinium* sp.	7.5	*Symbiodinium* sp.
	19.7	*Karlodinium veneficum*	5.6	*Scrippsiella* sp.	5.0	*Noctiluca scintillans*	1.5	*Noctiluca scintillans*
	0.9	*Cochlodinium* sp.	5.3	*Karlodinium* sp.	3.7	*Bysmatrum arenicola*	0.6	*Karlodinium* sp.
	0.9	*Pelagodinium* sp.	3.8	*Noctiluca* sp.	1.7	*Protoperidinium* sp.	0.6	*Alexandrium* sp.
	0.6	*Noctiluca scintillans*	3.5	*Neoceratium* sp.	0.9	*Karlodinium* sp.	0.6	*Peridinium* sp.
	0.3	*Akashiwo* sp.	1.3	*Heterocapsa* sp.	0.5	*Ceratium* sp.	0.6	*Amphidiniopsis* sp.
	0.2	*Paragymnodinium* sp.	1.1	*Blastodinium* sp.	0.4	*Erythropsidinium* sp.	0.5	*Heterocapsa triquetra*
	0.2	*Pfiesteria piscicida*	1.1	*Protodinium* sp.	0.4	*Akashiwo* sp.	0.4	*Cochlodinium* sp.
			1.0	*Chytriodinium* sp.	0.3	*Heterocapsa triquetra*	0.2	*Scrippsiella trochoidea*
Seong-san	29.1	*Gyrodinium* sp.	56.1	*Bysmatrum arenicola*	19.2	*Karlodinium veneficum*	22.4	*Gyrodinium* sp.
	16.6	*Gymnodinium* sp.	13.2	*Gyrodinium* sp.	11.1	*Bysmatrum arenicola*	19.3	*Karlodinium* sp.
	8.4	*Erythropsidinium* sp.	7.4	*Erythropsidinium* sp.	11.0	*Gyrodinium* sp.	13.9	*Bysmatrum arenicola*
	3.9	*Karlodinium* sp.	5.6	*Karlodinium* sp.	10.4	*Ceratium* sp.	6.8	*Ceratium* sp.
	1.7	*Heterocapsa* sp.	1.6	*Ceratium* sp.	8.2	*Peridinium* sp.	5.5	*Akashiwo* sp.
	0.7	*Paragymnodinium* sp.	1.3	*Pelagodinium* sp.	1217	*Akashiwo* sp.	2.3	*Heterocapsa* sp.
	0.3	*Azadinium* sp.	1.2	*Heterocapsa triquetra*	4.9	*Peridiniopsis* sp.	1.7	*Peridinium* sp.
	0.3	*Akashiwo* sp.	0.9	*Azadinium* sp.	4.0	*Gymnodinium catenatum*	1.5	*Azadinium* sp.
	0.2	*Cochlodinium* sp.	0.4	*Akashiwo* sp.	3.7	*Azadinium* sp.	1.1	*Noctiluca scintillans*
	0.2	*Pelagodinium* sp.	0.4	*Symbiodinium* sp.	2.6	*Heterocapsa circularisquama*	1.0	*Peridiniopsis* sp.

In Gunsan, the western coast, the highest number of reads detected by metagenomics data was seen in December (Table 1a). In March, two species (*Karlodinium veneficum* and *Gyrodinium* sp.) were dominant. When the ratio (%) of the top 10 species was calculated based on the total reads matched with dinoflagellate sequence, *Karlodinium veneficum* was the most dominant species, approximately 32%. Next, *Gyrodinium* sp. (24%) and *Gymnodinium* sp. (8%). In June, the composition of *Gonyaulax* sp. showed approximately 45%, followed by that of *Symbiodinium* sp. (15%) and *Karlodinium veneficum* (8%). Similar to

March, the dominant species in September was *Karlodinium* sp., which accounted for 24%. In December, *Gyrodinium* sp. (34%) and *Amphidiniella* sp. (25%) were dominant as well as *Ceratium* sp. which accounted for 11% (Table 1b, Figure 3a).

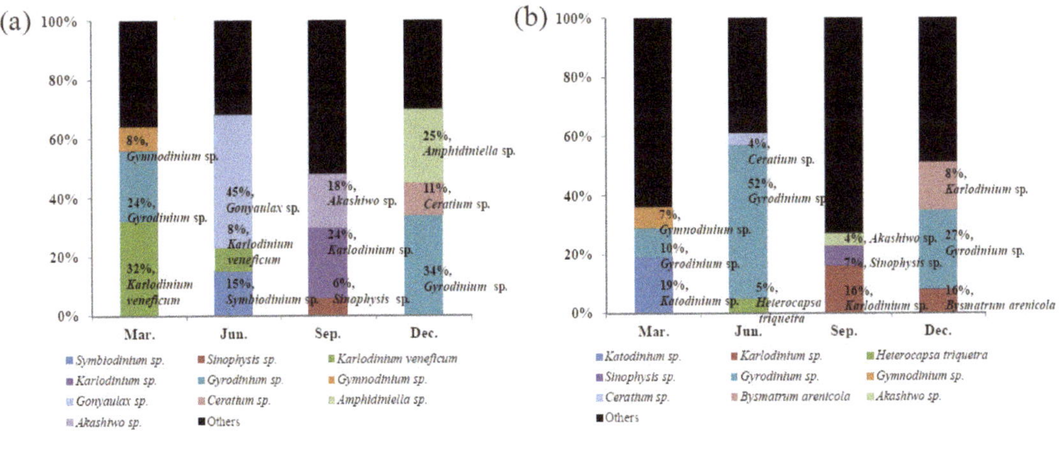

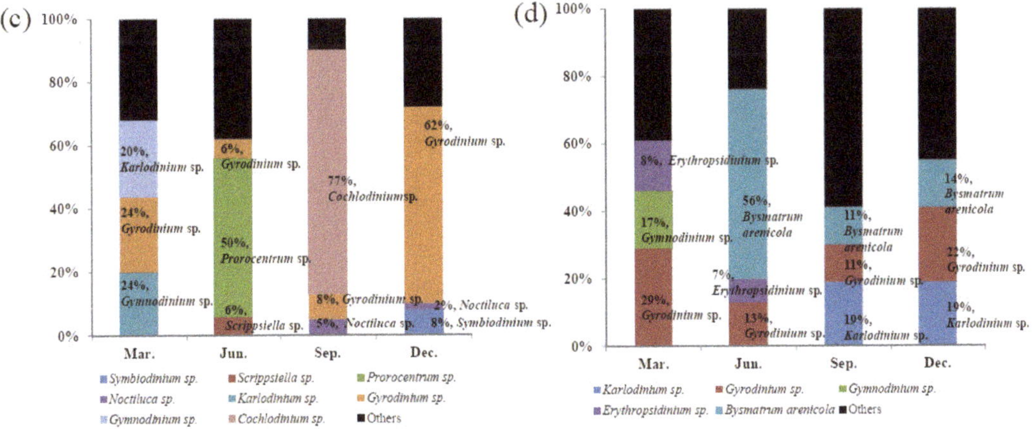

Figure 3. Proportion of top 3 most abundant species in each coastal seawater by metagenome analysis. Gunsan (**a**), Pohang (**b**), Tongyeong (**c**), Seongsan (**d**).

In Pohang, the eastern coast, *Gyrodinium* sp. was dominant in June and December, *Katodinium* sp. was dominant in March, and *Karlodinium* sp. was dominant in September. (Table 1, Figure 3b). In Tongyeong, the southern coast, the appearance of *Gyrodinium* sp. was high in March and December, and was dominant at 24% and 62%, respectively. In particular, *Cochlodinium* sp. formed a red tide and dominated over 77%, and in June, the dominance of *Prorocentrum* sp. was more than 50%. The diversity was the highest in December, when 28 species of dinoflagellate reads were detected (Table 1, Figure 3c). In Seongsan, *Gyrodinium* sp. appeared at a high rate in all seasons, while *Bysmatrum arenicola* and *Karlodinium veneficum* dominated in June (56%) and September (19%), respectively. The sand-dwelling dinoflagellate *Bysmatrum arenicola* was dominant at Seongsan, except in March (Table 1, Figure 3d).

Figure 4 illustrates the most common species in the four coastal waters. In March, there were three common species at all sampling sites: *Cochlodinium* sp., *Gyrodinium* sp. *Gymnodinium* sp., and *Pelagodinium* sp. The most common species in June were *Akashiwo* sp., *Karlodinium* sp., *Peridinium* sp., *Pelagodinium* sp., and *Prorocentrum* sp. In September, *Akashiwo* sp., *Bysmatrum* sp., *Ceratium* sp., *Katodinium* sp., *Sinophysis* sp., and *Peridinium* sp. were common. The common species in December were *Akashiwo* sp., *Alexandrium* sp., *Bysmatrum* sp., *Gyrodinium* sp., Hetrocapsa sp., *Peridiniopsis* sp., *Prorocentrum* sp., *Scrippsiella* sp., and *Symbiodinium* sp.

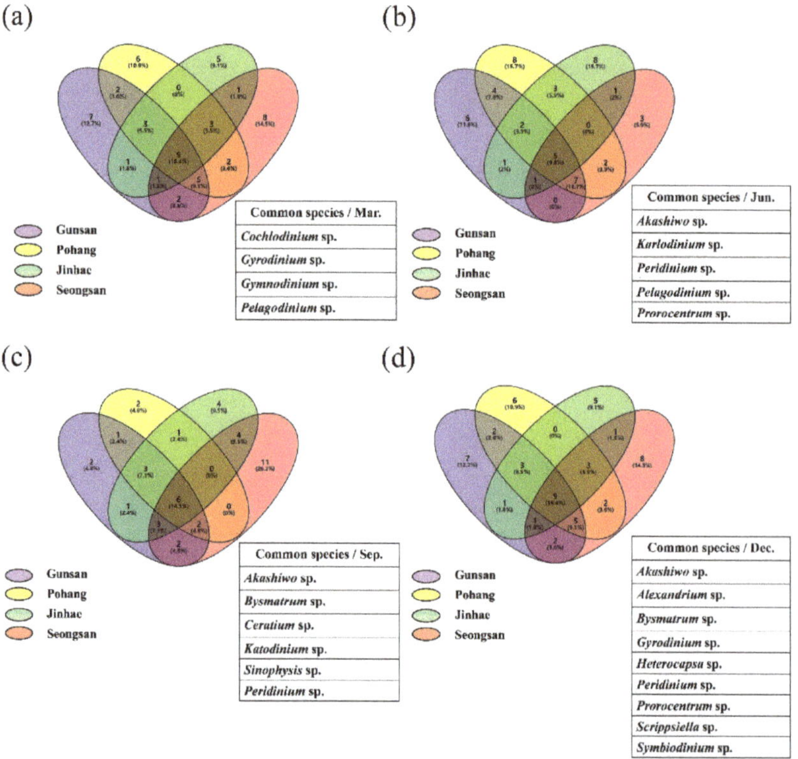

Figure 4. Seasonal common dinoflagellate species in 4 coastal waters by metagenome analysis. March (**a**), June (**b**), September (**c**), and December (**d**).

3.4. Comparison of Metagenomic Analysis and Microscopic Observation

When the abundance of dinoflagellates was analyzed by microscopic observation, the number of species composition was mostly lower than from metagenomic analysis (Table 2). Overall, the number of species in December was lower than in other seasons, as the biomass was considerably low and mainly dominated by diatoms. At Gunsan, the abundance of *Gyrodinium* sp. species was 0.8–2.9 cells mL^{-1} in March, September, and December, which showed similar patterns to the metagenomic analysis. In Pohang, the species composition in June was more diverse than in the other seasons, and two species of *Heterocapsa rotundata* (77.8 cells mL^{-1}) and *Heterocapsa triquetra* (12.1 cells mL^{-1}) were dominant. Similarly, the number of reads of *Heterocapsa triquetra* detected by the metagenomic analysis in the same sample were high. In Tongyeong, cell abundance of *Prorocentrum triestinum* (June) and *Cochlodinium polykrikoides* (September) was 341.1 and 2034 cells mL^{-1}, respectively, which was similar to the metagenome result that the number of reads of *Prorocentrum* sp. and *Cochlodinium* sp. was 11,335, and 53,412, respectively. Small thecated dinoflagellate species, such as *Azadinium* sp. and *Bysmatrum* sp., occurred in the Seongsan region, located at Jeju Island. Some small nano-planktonic dinoflagellates, which are difficult to identify by microscopy, were easily found at Seongsan and Tongyeong using the metagenomic analysis (Table 2).

Table 2. Species composition and cell number of dinoflagellates analyzed by microscopic observation. Seasonal (March, June, September, December) species composition in four coastal regions (Gunsan, Pohang, Tongyeong, Seongsan).

Location	March		June		September		December	
	Cell/mL	Species	Cell/mL	Species	Cell/mL	Species	Cell/mL	Species
Gunsan	2.7	*Heterocapsa triquetra*	2.7	*Scrippsiella* sp.	0.8	*Gyrodinium* sp.	2.9	*Gymnodinium* sp.
	0.9	*Gyrodinium* sp.	2.5	*Ceratium fusus*	0.1	*Peridiniopsis* sp.	2.9	*Gyrodinium* sp.
	0.9	*Prorocentrum micans*	2.5	*Heterocapsa rotundata*	0.1	*Protoperidinium divergence*		
	0.9	*Pyrocystis lunula*	1.8	*Gonyaulax* sp.				
			1.2	*Prorocentrum* sp.				
			0.9	*Dissodinium pseudolunula*				
			0.6	*Ceratium* sp.				
			0.6	*Ceratium tripos*				
			0.6	*Karlodinium* sp.				
			0.6	*Prorocentrum micans*				
Pohang	3.6	*Gymnodinium* sp.	77.8	*Heterocapsa rotundata*	1.7	*Heterocapsa rotundata*	0.4	*Gymnodinium* sp.
	1.8	*Gyrodinium* sp.	12.1	*Heterocapsa triquetra*	1.7	*Scrippsiella* sp.		
	1.4	*Ceratium kofoidii*	7.8	*Gymnodinium* sp.	0.8	*Prorocentrum triestinum*		
	1.2	*Alexandrium* sp.	4.3	*Protopeidinium pyriforme*	0.8	*Gymnodinium* sp.		
	0.5	*Heterocapsa rotundata*	2.6	*Gyrodinium* sp.	0.8	*Gyrodinium* sp.		
			2.6	*Ceratium kofoidii*				
			1.7	*Alexandrium* sp.				
			0.9	*Amphidinium operculatum*				
Tongyeong	1.6	*Small thecated dinoflagellate	341.1	*Prorocentrum triestinum*	2034	*Cochlodinium polykrikoides*	1.6	*Gymnodinium* sp.
	0.7	*Small naked dinoflagellate	18.0	*Small naked dinoflagellate	28.8	*Karlodinium* sp.		
	0.1	*Alexandrium* sp.	17.0	*Small thecated dinoflagellate	18.0	*Gyrodinium* sp.		
	0.1	*Gymnodinium* sp.	14.9	*Scrippsiella* sp.	5.4	*Prorocentrum* sp.		
	0.1	*Karlodinium* sp.	11.7	*Peridinium* sp.	3.6	*Bysmatrum* sp.		
			10.6	*Alexandrium* sp.	3.6	*Ceratium* sp.		
			3.2	*Heterocapsa* sp.	1.8	*Alexandrium* sp.		
			3.2	*Scrippsiella trochoidea*	1.8	*Heterocapsa* sp.		
			2.1	*Protoperidinium* sp.				
			1.1	*Gonyaulax* sp.				
			1.1	*Gymnodinium* sp.				
Seongsan	0.6	*Small naked dinoflagellate	5.8	*Azadinium* sp.	0.6	*Small naked dinoflagellate	1.3	*Bysmatrum* sp.
	0.2	*Bysmatrum* sp.	5.8	*Bysmatrum* sp.	0.5	*Peridiniopsis* sp.	1.0	*Gymnodinium* sp.
	0.2	*Katodinium* sp.	5.4	*Small naked dinoflagellate	0.3	*Gymnodinium* sp.	0.3	*Gyrodinium* sp.
	0.2	*Prorocentrum* sp.	2.4	*Small thecated dinoflagellate	0.3	*Prorocentrum minimum*		
			1.4	*Gymnodinium* sp.				
			1.0	*Protoperidinium pellucidum*				
			0.3	*Heterocapsa* sp.				
			0.3	*Peridiniopsis* sp.				
			0.3	*Prorocentrum* sp.				
			0.3	*Protoperidinium* sp.				
			0.3	*Woloszynskia* sp.				

*Small thecated dinoflagellates: *Apicoporus, Azadinium, Crypthecodinium, Durinskia, Heterocapsa, Pfiesteria*. *Small naked dinoflagellates: *Amphidiniopsis, Biecheleria, Karlodinium, Gymnodinium, Gyrodiniellum, Paragymnodinium, Pelagodinium, Symbidinium*.

Although not all species of dinoflagellates identified by microscopic observation were included in the metagenomic analysis, the appearance of dominant species was found to be quite similar (Table 2).

3.5. Seasonal Distribution of Harmful Species Based on Metagenomic Analysis

Four species of dinoflagellates (*Cochlodinium* sp., *Alexandrium* spp., *Dinophysis* spp., and *Gymnodinium* sp.) were selected as the causative species of red tide formation or toxin production in Korean waters (Figure 5a), and their seasonal distribution characteristics based on the number of reads through metagenomic analysis was confirmed by region. In Gunsan, the reads of *Gymnodinium* sp. were considerable in March, and *Dinophysis* spp. appeared in June and September. In Pohang, *Gymnodinium* sp. was relatively high in March, and *Cochlodinium* sp. was also detected at a high distribution in December. In Tongyeong, the abundance of *Cochlodinium* sp. was especially high in September, when a red tide from this species was occurring. In Seongsan, the appearance of *Gymnodinium* sp. in March and September was revealed by microscopic observation (Figure 5b). Based on these findings, the seasonal distribution of red tide-causing species, which was not confirmed by microscopic observation, was confirmed using metagenomic analysis.

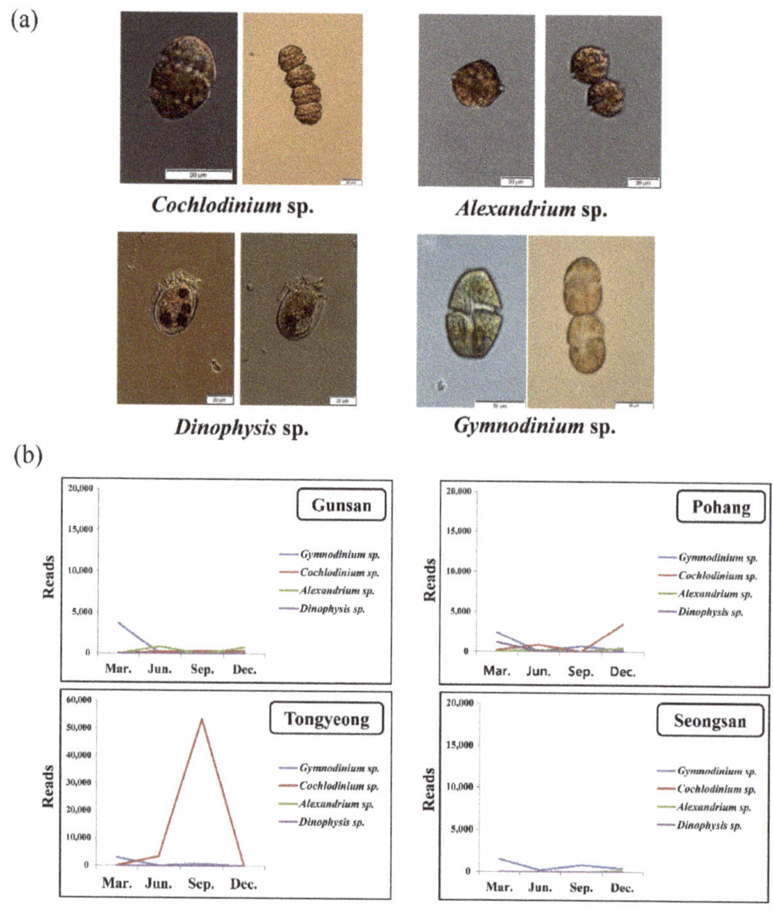

Figure 5. Seasonal distribution of red-tide-causing species through metagenome analysis. Photo of red-tide-causing species (*Cochlodinium* sp., *Alexandrium* sp., *Dinophysis* sp., *Gymnodinium* sp.) taken under a light microscope (**a**), seasonal changes in red-tide-causing species (**b**).

4. Discussion

Approximately 300 dinoflagellate species are known to cause red tides and produce toxins worldwide, and these harmful events are increasing with changes in human activities and the environment [4]. Toxic dinoflagellate blooms frequently occur in the southern coastal waters of Korea, where many cage fish farms are located. As shown in Table 1, *Cochlodinium* sp. were dominant at Tongyeong in September according to NGS, which corresponds to the cell abundance counted by microscopic observation. In June, the NGS result that *Prorocentrum* sp. were mainly observed at Tongyeong was similar to the occurrence detected by microscopy analysis at this location. In addition, *Karlodinium* sp., which produces Karlotoxin and induces hemolytic and cytotoxic activity associated with fish mortality, appeared in our NGS results [27].

In a situation where the morphological analysis method is the dominant method for diagnosing harmful dinoflagellates off Korean coasts, diagnosis using molecular biology is considered to be a more objective number, and the development of technology through this method can lead to the development of new monitoring techniques [28]. Moreover, if NGS technology has been developed and applied to the monitoring of marine organisms, it is possible to simultaneously analyze a large amount of mixed samples and save the effort and time of long-term monitoring and research analysis [29–31].

Monitoring of marine microalgae using NGS has been used by many researchers because of its various advantages [11]. Metagenomic analysis using NGS has revealed a significant number of phytoplankton taxa previously missed by microscopy in recent efforts to sequence marine microorganisms [32]. Our study also revealed a significant number of dinoflagellate communities that could not be distinguished microscopically. The genetic analysis method used in this study, especially high-throughput sequencing, has shown effectiveness in the study of phytoplankton diversity and ecology, and it is considered that it can potentially replace the microscopic identification and population quantification methods currently used.

Light microscopy, which has been used for morphological classification and population evaluation, requires an extensive amount of consideration. Underestimation of phytoplankton, including dinoflagellates, in microscopic samples results in cell loss of taxa during preservation, storage, and handling, preferentially after treatment of samples with fixing fluid. Further, when counting cells, a sedimentation chamber is commonly used, which means that smaller cells that do not sink sufficiently are less counted or missed [33]. Moreover, identification of small dinoflagellates using microscopy is not easy when their cell size is under 20 μm with similar morphologies when fixed with Lugol's solution [34]. We found that a significant number of dinoflagellate species were confirmed by metagenomic analysis compared to that by microscopic analysis. The small dinoflagellate cells which were classified as 'small naked dinoflagellate' were positively identified as species belonging to the genera *Amphidiniopsis, Biecheleria, Gymnodinium, Gyrodiniellum, Paragymnodinium, Pelagodinium,* and *Symbidinium,* while 'small thecated dinoflagellate' included *Apicoporus, Azadinium, Crypthecodinium, Durinskia, Heterocapsa,* and *Pfiesteria.* In particular, the sand-dwelling dinoflagellate *Bysmatrum arenicola,* which is easily confused with *Scrippsiella* [35] in microscopic analysis, was found in the metagenomic analysis in June at Seongsan (Figure 3d). This suggests that the metagenomic analysis was more extensive.

Although the NGS technique showed a high resolution for species identification compared to that with conventional microscopic analysis, further studies are required for development of an understanding of the spatial and seasonal dynamics of the dinoflagellate community using NGS-based metagenomics. Thus far, molecular markers based on ribosomal DNA have usually been used to identify the species, even among relatives [36]. However, this approach is limited by interspecific divergence, while it is difficult to distinguish intraspecific variation. As the reference database of dinoflagellates via the NGS method in this study was established based on 18S rDNA sequences, the relative proportions of some dinoflagellates in field samples could be misidentified in the presence of other dinoflagellates which were similar. Large subunit (LSU) rDNA sequences of *Prorocentrum* species containing *P. rhathymum, P. mexicanum,* and *P.* cf. *rhathymum,* which are toxic, were closer to the relatives, showing 0.1–0.9% dissimilarity, and small subunit rDNA (SSU) sequences of most of these are nearly identical [37]. Edvardsen et al. [38] reported that SSU rDNA sequences among *Dinophysis acuminata, Dinophysis acuta,* and *Dinophysis norvegica* show approximately 0.3% distance, and differences of LSU rDNA sequences among these species show 0.4–1.6% distance. Moreover, species whose sequences are not available in the GenBank are hardly detected despite their potential presence in the sample analyzed by the NGS technique because of the absence of deposited sequences. To distinguish intraspecific similarity of the above-mentioned species, establishment of a reference database via the NGS technique based on biomarkers such as cytochrome c oxidase I (COX1) and the cytochrome b (COB) gene which allows for the unambiguous identification of the species should be developed.

Metagenomic analysis of marine biodiversity and abundance based on NGS will provide precise indicators for understanding biological patterns and characteristics of species in different habitats. Given the lack of molecular reference library databases, it is necessary to collect vast amounts of sequence information targeting biomarkers such as SSU, LSU, COX1, and COB genes. However, in this study, we established a reference database of dinoflagellates that occur in the coastal waters of Korea based on SSU rDNA sequences using the NGS technique and analyzed field samples in the presence of this NGS reference database library. We expect that the newly established reference database via the NGS will provide a better understanding of the seasonal dynamics of toxic dinoflagellates, as well as a complementary approach to conventional microscopic analysis for monitoring dinoflagellate community compositions.

5. Conclusions

This study integrated analyses of high-resolution dinoflagellate community composition and distribution in South Korea. Altogether, the results presented here reveal a complex dinoflagellate community pattern. The NGS-based (18S rRNA amplicon) metagenomics were able to detect dinoflagellates with low abundance, and allow continuous monitoring of the phytoplankton community in environmental samples even though numerous DNA samples were simultaneously collected compared to the conventional microscopic analysis. Our analysis suggested that NGS-based characterization of the 18S rRNA gene holds great promise as a tool for phytoplankton monitoring, as it allows for simultaneous regional cluster analysis monitoring in a high-throughput, reproducible, and cost-effective manner.

In today's world, which requires advances in environmental monitoring due to large-scale blooming of toxic algae and international regulations regarding their toxic substances, this study provides a technique for the rapid evaluation of environmental samples for existing taxa of major dinoflagellates and potentially harmful/invasive species. In addition, the extension of the reference database presented in this study and addition of the species list can further expand the taxonomic scope so it can be applied to real-time monitoring of temporal dynamics and species diversity problems of harmful algal blooms in a wide range of waters.

Supplementary Materials: The following supporting information can be downloaded at: https://www.mdpi.com/article/10.3390/microorganisms10071459/s1, Figure S1: Workflow chart for metagenome analysis, Using CLC Genomics Workbench program. Supplementary Table S1.

Author Contributions: Conceptualization, J.H., H.W.K. and. J.P.; field investigation, S.J.M., J.-H.H., and E.S.L.; software, J.H.; validation and formal analysis, J.H. and J.-H.H.; data curation, J.P; writing—original draft preparation, J.H.; writing—review and editing, J.H. and J.P. All authors have read and agreed to the published version of the manuscript.

Funding: This study was supported by the project 'Development of aquaculture technology for pomfrets as an endemic fish species on West Sea' (R2022005) of National Institute of Fisheries Science (NIFS), Incheon, South Korea, and supported by the National Research Foundation of Korea (NRF) grant funded by the Korea government (MSIT)(NRF-2021R1A2C1005943).

Data Availability Statement: Not applicable.

Acknowledgments: We thank E.J.K. for field sampling. We thank the reviewers for their comments.

Conflicts of Interest: The authors declare no conflict of interest.

References

1. Lessard, E.J. *Oceanic Heterotrophic Dinoflagellates: Distribution, Abundance, and Role as Microzooplankton*; Rhode Island University: Kingston, NY, USA, 1984.
2. Jeong, H.J.; Yoo, Y.D.; Kim, J.S.; Seong, K.A.; Kang, N.S.; Kim, T.H. Growth, feeding and ecological roles of the mixotrophic and heterotrophic dinoflagellates in marine planktonic food webs. *Ocean Sci. J.* **2010**, *45*, 65–91. [CrossRef]

3. Kwon, H.-K.; Ortiz, L.C.; Vukasin, G.D.; Chen, Y.; Shin, D.D.; Kenny, T.W. An oven-controlled MEMS oscillator (OCMO) with sub 10 MW, ±1.5 ppb stability over temperature. In Proceedings of the 2019 20th International Conference on Solid-State Sensors, Actuators and Microsystems & Eurosensors XXXIII (TRANSDUCERS & EUROSENSORS XXXIII), Berlin, Germany, 23–27 June 2019; pp. 2072–2075.

4. Zohdi, E.; Abbaspour, M. Harmful algal blooms (red tide): A review of causes, impacts and approaches to monitoring and prediction. *Int. J. Environ. Sci. Technol.* **2019**, *16*, 1789–1806. [CrossRef]

5. Couet, D.; Pringault, O.; Bancon-Montigny, C.; Briant, N.; Poulichet, F.E.; Delpoux, S.; Yahia, O.K.-D.; Hela, B.; Herve, F.; Rovillon, G. Effects of copper and butyltin compounds on the growth, photosynthetic activity and toxin production of two HAB dinoflagellates: The planktonic *Alexandrium catenella* and the benthic *Ostreopsis* cf. *ovata*. *Aquat. Toxicol.* **2018**, *196*, 154–167. [CrossRef] [PubMed]

6. Chakraborty, S.; Pančić, M.; Andersen, K.H.; Kiørboe, T. The cost of toxin production in phytoplankton: The case of PST producing dinoflagellates. *ISME J.* **2019**, *13*, 64–75. [CrossRef]

7. Jantschke, A.; Pinkas, I.; Schertel, A.; Addadi, L.; Weiner, S. Biomineralization pathways in calcifying dinoflagellates: Uptake, storage in MgCaP-rich bodies and formation of the shell. *Acta Biomater.* **2020**, *102*, 427–439. [CrossRef]

8. Choi, H.; Liu, X.; Kim, H.I.; Kim, D.; Park, T.; Song, S. A Facile Surface Passivation Enables Thermally Stable and Efficient Planar Perovskite Solar Cells Using a Novel IDTT-Based Small Molecule Additive. *Adv. Energy Mater.* **2021**, *11*, 2003829. [CrossRef]

9. Chan, K.K.; Tan, T.J.; Narayanan, K.K.; Procko, E. An engineered decoy receptor for SARS-CoV-2 broadly binds protein S sequence variants. *Sci. Adv.* **2021**, *7*, eabf1738. [CrossRef]

10. Lee, E.S.; Kim, E.-K.; Choi, Y.-H.; Jung, Y.H.; Kim, S.Y.; Koh, J.W.; Choi, E.K.; Cheon, J.-E.; Kim, H.-S. Factors associated with neurodevelopment in preterm infants with systematic inflammation. *BMC Pediatrics* **2021**, *21*, 114. [CrossRef]

11. Muhammad, B.L.; Kim, T.; Ki, J.-S. 18S rRNA analysis reveals high diversity of phytoplankton with emphasis on a naked Dinoflagellate *Gymnodinium* sp. at the Han river (Korea). *Diversity* **2021**, *13*, 73. [CrossRef]

12. Akbar, M.A.; Sang, J.; Khan, A.A.; Amin, F.-E.; Hussain, S.; Sohail, M.K.; Xiang, H.; Cai, B. Statistical analysis of the effects of heavyweight and lightweight methodologies on the six-pointed star model. *IEEE Access* **2018**, *6*, 8066–8079. [CrossRef]

13. Gran-Stadniczeñko, S.; Egge, E.; Hostyeva, V.; Logares, R.; Eikrem, W.; Edvardsen, B. Protist diversity and seasonal dynamics in Skagerrak plankton communities as revealed by metabarcoding and microscopy. *J. Eukaryot. Microbiol.* **2019**, *66*, 494–513. [CrossRef] [PubMed]

14. Caporaso, J.G.; Kuczynski, J.; Stombaugh, J.; Bittinger, K.; Bushman, F.D.; Costello, E.K.; Fierer, N.; Peña, A.G.; Goodrich, J.K.; Gordon, J.I.; et al. QIIME allows analysis of high-throughput community sequencing data. *Nat. Methods* **2010**, *7*, 335–336. [CrossRef]

15. Kuczynski, J.; Stombaugh, J.; Walters, W.A.; González, A.; Caporaso, J.G.; Knight, R. Using QIIME to analyze 16S rRNA gene sequences from microbial communities. *Curr. Protoc. Microbiol.* **2012**, *S27*, 1E.5.1–1E.5.20. [CrossRef] [PubMed]

16. Osman, E.O.; Suggett, D.J.; Voolstra, C.R.; Pettay, D.T.; Clark, D.R.; Pogoreutz, C.; Sampayo, E.M.; Warner, M.E.; Smith, D.J. Coral microbiome composition along the northern Red Sea suggests high plasticity of bacterial and specificity of endosymbiotic dinoflagellate communities. *Microbiome* **2020**, *8*, 1–16.

17. Jaeckisch, N.; Yang, I.; Wohlrab, S.; Glöckner, G.; Kroymann, J.; Vogel, H.; Cembella, A.; John, U. Comparative genomic and transcriptomic characterization of the toxigenic marine dinoflagellate *Alexandrium ostenfeldii*. *PLoS ONE* **2011**, *6*, e28012. [CrossRef]

18. Eichholz, K.; Beszteri, B.; John, U. Putative monofunctional type I polyketide synthase units: A dinoflagellate-specific feature? *PLoS ONE* **2012**, *7*, e48624. [CrossRef] [PubMed]

19. Price, D.C.; Farinholt, N.; Gates, C.; Shumaker, A.; Wagner, N.E.; Bienfang, P.; Bhattacharya, D. Analysis of *G. ambierdiscus* transcriptome data supports ancient origins of mixotrophic pathways in dinoflagellates. *Environ. Microbiol.* **2016**, *18*, 4501–4510. [CrossRef]

20. Liu, H.; Stephens, T.G.; González-Pech, R.A.; Beltran, V.H.; Lapeyre, B.; Bongaerts, P.; Cooke, I.; Aranda, M.; Bourne, D.G.; Forêt, S.; et al. Symbiodinium genomes reveal adaptive evolution of functions related to coral-dinoflagellate symbiosis. *Commun. Biol.* **2018**, *1*, 95. [CrossRef]

21. Lim, A.S.; Jeong, H.J. Benthic dinoflagellates in Korean waters. *Algae* **2021**, *36*, 91–109. [CrossRef]

22. Kim, H.S.; Yih, W.; Kim, J.H.; Myung, G.; Jeong, H.J. Abundance of epiphytic dinoflagellates from coastal waters off Jeju Island, Korea during autumn 2009. *Ocean Sci. J.* **2021**, *46*, 205–209. [CrossRef]

23. Kim, H.S.; Kim, S.H.; Jung, M.M.; Lee, J.B. New record of dinoflagellates around Jeju Island. *J. Ecol. Environ.* **2013**, *36*, 273–291. [CrossRef]

24. Shah, M.; Rahman, M.; An, S.J.; Lee, J.B. Seasonal abundance of epiphytic dinoflagellates around coastal waters of Jeju Island, Korea. *J. Mar. Sci. Technol.* **2013**, *21*, 20.

25. Stoeck, T.; Bass, D.; Nebel, M.; Christen, R.; Jones, M.D.; Breiner, H.W.; Richards, T.A. Multiple marker parallel tag environmental DNA sequencing reveals a highly complex eukaryotic community in marine anoxic water. *Mol. Ecol.* **2010**, *19*, 21–31. [CrossRef] [PubMed]

26. Gómez, F. A list of free-living dinoflagellate species in the world's oceans. *Acta Bot. Croat.* **2005**, *64*, 129–212.

27. Llanos-Rivera, A.; Álvarez-Muñoz, K.; Astuya-Villalón, A.; López-Rosales, L.; García-Camacho, F.; Sánchez-Mirón, A.; Rodriguez, J. Sublethal Effects of the Toxic Alga *Karlodinium veneficum* on Fish. *Res. Sq.* **2021**, *1*, e1021203.

28. Elleuch, J.; Ben Amor, F.; Barkallah, M.; Haj Salah, J.; Smith, K.F.; Aleya, L.; Fendri, I.; Abdelkafi, S. q-PCR-based assay for the toxic dinoflagellate *Karenia selliformis* monitoring along the Tunisian coasts. *Environ. Sci. Pollut. Res.* **2021**, *28*, 57486–57498. [CrossRef]
29. Abad, D.; Albaina, A.; Aguirre, M.; Laza-Martínez, A.; Uriarte, I.; Iriarte, A.; Villate, F.; Estonba, A. Is metabarcoding suitable for estuarine plankton monitoring? A comparative study with microscopy. *Mar. Biol.* **2016**, *163*, 149. [CrossRef]
30. Metfies, K.; von Appen, W.-J.; Kilias, E.; Nicolaus, A.; Nöthig, E.-M. Biogeography and photosynthetic biomass of arctic marine pico-eukaroytes during summer of the record sea ice minimum 2012. *PLoS ONE* **2016**, *11*, e0148512. [CrossRef]
31. Langer, J.A.; Sharma, R.; Schmidt, S.I.; Bahrdt, S.; Horn, H.G.; Algueró-Muñiz, M.; Nam, B.; Achterberg, E.P.; Riebesell, U.; Boersma, M. Community barcoding reveals little effect of ocean acidification on the composition of coastal plankton communities: Evidence from a long-term mesocosm study in the Gullmar Fjord, Skagerrak. *PLoS ONE* **2017**, *12*, e0175808. [CrossRef]
32. Chai, Z.Y.; He, Z.L.; Deng, Y.Y.; Yang, Y.F.; Tang, Y.Z. Cultivation of seaweed *Gracilaria lemaneiformis* enhanced biodiversity in a eukaryotic plankton community as revealed via metagenomic analyses. *Mol. Ecol.* **2018**, *27*, 1081–1093. [CrossRef]
33. Loeffler, C.R.; Richlen, M.L.; Brandt, M.E.; Smith, T.B. Effects of grazing, nutrients, and depth on the ciguatera-causing dinoflagellate *Gambierdiscus* in the US Virgin Islands. *Mar. Ecol. Prog. Ser.* **2015**, *531*, 91–104. [CrossRef]
34. Lee, S.Y.; Jeong, H.J.; Ok, J.H.; Kang, H.C.; You, J.H. Spatial-temporal distributions of the newly described mixotrophic dinoflagellate *Gymnodinium smaydae* in Korean coastal waters. *Algae* **2020**, *35*, 225–236. [CrossRef]
35. Horiguchi, T.; Pienaar, R.N. Validation of *Bysmatrum arenicola* horiguchi et pienaar, sp. nov.(dinophyceae)(Note). *J. Phycol.* **2000**, *36*, 237. [CrossRef]
36. Shearer, T.L.; Coffroth, M.A. DNA BARCODING: Barcoding corals: Limited by interspecific divergence, not intraspecific variation. *Mol. Ecol. Resour.* **2008**, *8*, 247–255. [CrossRef] [PubMed]
37. Lim, A.S.; Jeong, H.J.; Jang, T.Y.; Kang, N.S.; Lee, S.Y.; Yoo, Y.D.; Kim, H.S. Morphology and molecular characterization of the epiphytic dinoflagellate *Prorocentrum* cf. *rhathymum* in temperate waters off Jeju Island, Korea. *Ocean Sci. J.* **2013**, *48*, 1–17. [CrossRef]
38. Edvardsen, B.; Shalchian-Tabrizi, K.; Jakobsen, K.S.; Medlin, L.K.; Dahl, E.; Brubak, S.; Paasche, E. Genetic variability and molecular phylogeny of dinophysis species (dinophyceae) from norwegian waters inferred from single cell analyses of rDNA 1. *J. Phycol.* **2003**, *39*, 395–408. [CrossRef]

microorganisms

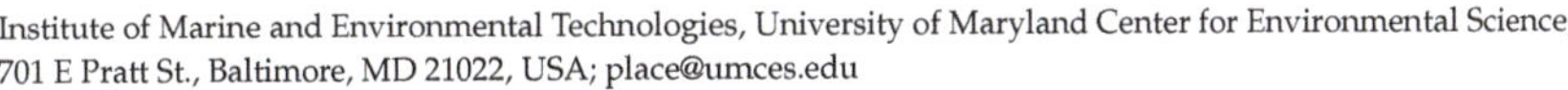

Article

A Strategy for Gene Knockdown in Dinoflagellates

Miranda Judd * and Allen R. Place

Institute of Marine and Environmental Technologies, University of Maryland Center for Environmental Science, 701 E Pratt St., Baltimore, MD 21022, USA; place@umces.edu
* Correspondence: mjudd@umces.edu; Tel.: +1-410-234-8828

Abstract: Dinoflagellates are unicellular protists that display unusual nuclear features such as large genomes, condensed chromosomes and multiple gene copies organized as tandem gene arrays. Genetic regulation is believed to be controlled at the translational rather than transcriptional level. An important player in this process is initiation factor eIF4E which binds the 7-methylguanosine cap structure (m7G) at the 5′-end of mRNA. Transcriptome analysis of eleven dinoflagellate species has established that each species encodes between eight to fifteen eIF4E family members. Determining the role of eIF4E family members in gene expression requires a method of knocking down their expression. In other eukaryotes this can be accomplished using translational blocking morpholinos that bind to complementary strands of RNA, therefore inhibiting the mRNA processing. Previously, unmodified morpholinos lacked the ability to pass through cell membranes, however peptide-based reagents have been used to deliver substances into the cytosol of cells by an endocytosis-mediated process without damaging the cell membrane. We have successfully delivered fluorescently-tagged morpholinos to the cytosol of *Amphidinium carterae* by using a specific cell penetrating peptide with the goal to target an eIF4e-1a sequence to inhibit translation. Specific eIF4e knockdown success (up to 42%) has been characterized via microscopy and western blot analysis.

Keywords: dinoflagellate; knockdown; morpholino; translation

Citation: Judd, M.; Place, A.R. A Strategy for Gene Knockdown in Dinoflagellates. *Microorganisms* **2022**, *10*, 1131. https://doi.org/10.3390/microorganisms10061131

Academic Editors: Shauna Murray and Thomas Mock

Received: 1 April 2022
Accepted: 27 May 2022
Published: 31 May 2022

Publisher's Note: MDPI stays neutral with regard to jurisdictional claims in published maps and institutional affiliations.

1. Introduction

Dinoflagellates are single-celled eukaryotes and members of the Alveolate lineage [1]. Dinoflagellates exhibit extremely diverse trophic strategies, including predation, photo-autotrophy, mixotrophy, and intracellular parasitism [2,3]. Most cultured dinoflagellate species are photosynthetic, making them key marine primary producers. They are well-known for bloom formation in coastal waters, making toxins that bioaccumulate in the food chain, producing bioluminescence, and as coral symbionts [4–6].

Climate-change has caused a warming of the Earth's oceans, benefitting the formation of harmful algal blooms [7]. Of the algal species that have been reported as producing marine harmful blooms, 75% are dinoflagellates [8–11]. Accumulation of dinoflagellates in coastal waters has begun to increase the presence of red tides, bringing with it fish mass mortality and marine toxin-derived disease in humans [12]. Increasing water temperatures provide optimum growth conditions for many dinoflagellates, allowing for increased toxic effects on their environment [13,14]. Globally, previous research has confirmed the mechanism and structure of some of these toxins [15–20].

A hallmark of these toxic blooms can be traced to the production of complex secondary metabolites. Some of these toxins are thought to assist in prey capture through the formation of a nonspecific pore upon complexation with prey's sterol membrane components [16,21]. Unfortunately, research into the biosynthesis of these dinoflagellate toxins is sorely lacking. This is in large part due to dinoflagellates having unusual cell biology [22]. Their genomes are larger than typical protists, with about $1.2–112 \times 10^9$ base pairs of DNA per haploid genome [10], whereas other protist genomes range in the millibases [23,24]. Dinoflagellate chromosomes are condensed into liquid crystalline states throughout the cell cycle and lack nucleosomes, instead using histone-like proteins (HLPs) that are more similar to bacterial DNA binding proteins. Many dinoflagellate genes are organized in multiple copies as tandem repeats, some of which may be present in up to ~10^5 copies. Increasingly transcriptomic data has shown that dinoflagellates express numerous genes, yet about 50% have no match to known sequences [25]. The function of these sequences, as well as the effects of identified sequences, still need to be established through functional genomic studies.

Control of post-transcriptional regulation in dinoflagellates is currently enigmatic, with mRNA levels showing no correlation to protein production. Because dinoflagellates are believed to regulate at the translational level, rather than during transcription, translation factors are of great interest when understanding dinoflagellate metabolism. In most known eukaryotic translation systems, eIF4Es function as a rate-limiting step toward protein synthesis [26,27]. eIF4E is part of an extended gene family found exclusively in eukaryotes. This translation factor binds to the mRNA cap to recruit the ribosome for translation initiation. In most studied eukaryotic systems (excluding plants), the eIF4E-1 family member is expressed ubiquitously in all cell types from a single copy, such as in *Homo sapiens* or *Saccharomyces cerevisiae* [28–32]. Early studies speculated that eukaryotic systems contain a single gene that encodes eIF4E [33], but since then sequencing projects have revealed that many organisms contain multiple genes encoding proteins that have sequence similarity to the recognized eIF4E [31,34–36]. In the case of dinoflagellates, genes regularly appear in multiple copies, with eIF4E being no exception [37,38]. These gene copies commonly appear as slightly different variants with distinctive degrees of diversity. Prior transcriptome analysis of eleven dinoflagellate species has established that each species encodes between eight to fifteen eIF4E family members, a number surpassing that found in any other eukaryotes, including other alveolates [22].

Core dinoflagellate eIF4E translation factors are divided into 3 clades (1, 2, and 3), along with 3 subclades within each (a, b, c); with a total of 9 members. Our previous work has shown that these eIF4E family members display divergences at critical amino acids, suggesting the family members are functionally distinct [22,39]. Of these 3 major clades, eIF4E-1 stands out as the most duplicated, and with the lowest number of substitutions. Based on the expression levels of the subclades, our lab has theorized that eIF4E-1a is likely the primary translation initiation factor.

Although expression of subclade eIF4E-1a is highest of all the family members, dinoflagellates still generate a greater diversity and degree of eIF4E duplications than seen in other eukaryotes [22]. Understanding their various roles will bring us closer to understanding how dinoflagellates adapt to their environment, giving insight into harmful algal bloom formations, as well as their production of complex secondary metabolites and toxin biosynthesis. Generally, eIF4E family members are known to have different roles in metazoan gene expression [40]. Similarly, we predict that dinoflagellate eIF4Es will have distinct functions, allowing for an increased dependence on the translational control of gene expression [22]. Determining the role of the eIF4E family members requires a method of knocking down their expression. The unusual cell biology of dinoflagellates makes common gene knockout strategies impractical, limiting the amount of genetic research that can be applied. This is because gene knockouts require a "deletion" of all operable gene-copies, which can be difficult to obtain when many copies with slight variations exist [38]. In this case, gene knockdown strategies are the most feasible, as they target the functional transcripts produced by the gene copies, which in many cases remain less diverse [39,41].

In particular, we are pursuing the use of an antisense-based knockdown approach in order to study how a decrease in target gene expression effects dinoflagellate metabolism. In prior research, the introduction of antisense-oligomers to dinoflagellate cells has been hampered by their thick, cellulosic cell wall [42,43]. Other studies have bypassed this obstacle by preparing spheroplasts, cells with a completely or partially removed cell-wall, beforehand [42–44]. Spheroplast production is done by incubating cells on plates in a polyethylene glycol (PEG) solution, which promotes fusion of the vesicles and cell membrane, and ultimately a decrease in total cellulose. So far only two studies have been successful in achieving gene knockdown with this spheroplast procedure; targeting a condensin subunit and targeting a cellulose synthase [42–44]. Once introduced, the antisense-oligo was able to bind to cytoplasmic mRNA and knockdown expression of the target gene. Although gene expression could be quantified in this way, it appears that some physiological effects were hidden by the effects of PEG on the cell wall, which causes the cells to lose rigidity. Also, the need for cell plating, rather than cell culturing, immensely limits the species of dinoflagellates that can be studied since many will not grow outside of a liquid medium.

There has also been evidence of RNA interference (RNAi) machinery within dinoflagellates, a naturally occurring mechanism for gene silencing through various methods such as RNA degradation, transcriptional repression and translation inhibition [25,45,46]. One study observed the effects of RNAi silencing tool on the proton-pump rhodopsin and CO_2-fixing enzyme Rubisco encoding genes in dinoflagellates by introducing small interfering RNAs (siRNAs) to dinoflagellate cultures via immersion. Results showed success in gene suppression within the two dinoflagellate species studied, *Prorocentrum donghaiense* and *Karlodinium veneficum* [25]. This decrease in gene expression was observed with a decrease on overall growth rate for both species as well, compared to the control green fluorescent protein (GFP) labelled siRNA. This knockdown method in dinoflagellates was initially challenging due to the large copy number of target DNA and permanently-condensed chromatin, but recent research has shown that there is a strong possibility that knockdown procedures can be successful.

Recently our team has begun to develop a system for introducing antisense morpholinos into dinoflagellate cells without the use of PEG to warp the cell wall, or the use of RNAi. Instead, we are using a novel delivery peptide that delivers substances via an endocytosis-mediated process that avoids damaging the plasma membrane of the cell [47]. Not only this, but the knockdown process can be done completely by immersion. The peptide and antisense morpholino are added directly to the culture to stimulate endocytosis and morpholino uptake. To test this system on dinoflagellates, we used the manufacturer recommended concentrations of delivery-peptide and antisense-morpholino on a dense culture of *Amphidinium carterae*, a known algal bloom species. The antisense morpholino is targeted to what we believed to be the main dinoflagellate translation factor, eIF4E-1a [22,27,40,48]. Our preliminary data has shown that unlike with PEG addition, the delivery peptide does not cause the cell population to drastically decrease.

2. Materials and Methods

2.1. Cell Culturing

Amphidinium carterae (Hulbert) strain CCMP1314 was grown in ESAW artificial marine media with a salinity of 32 ppt supplemented with f/2 nutrients without silicates at 25 °C [49]. The medium was buffered with 1mM HEPES (pH 8.0). Since bacterized cultures have shown to affect analyses of translation rate in *A. carterae*, the cultures were maintained axenically with an antibiotic solution of kanamycin (50 µg/mL), carbenicillin (100 µg/mL), and spectinomycin (50 µg/mL) [50]. The cultures were grown under 100 µmol/m^2 s$^-$ light. Delivery of the morpholinos (described below) requires constant swirling to keep the reagents in solution, therefore the cultures were also placed on an orbital shaker at 60 rpm [47]. The cultures were allowed to acclimate to the swirling for a week before knockdown reagents were added.

2.2. Morpholino Customization and Delivery

The sequence of the initiation factor eIF4E-1a was found by our lab previously [22]. Morpholino antisense oligonucleotides (MOs) are nucleic acid analogues in which DNA bases are bound to a non-charged backbone (morpholine rings linked by phosphorodiamidate bonds) [51]. For our purposes, a translation-blocking MO was created that covered the eIF4E-1a translational start site (5′-TCATTGAAGCTCAAACAAGCCATTG-3′). Specificity for the intended target sites was verified by BLAST analysis against the *Amphidinium carterae* transcriptome. MOs were purchased from GeneTools (Philomath, OR, USA) and modified with a red-emitting fluorescent 3′ Lissamine addition and then used at a concentration of 1 μM and 10 μM. Standard control MOs with the Lissamine addition were ordered as well (5′-CCTCTTACCTCAGTTACAATTTATA-3′).

MOs were delivered using Endo-Porter reagent (GeneTools, Philomath, OR, USA) at a concentration of 4 μM. Cultures of *A. carterae* seeded in 12-well plates were treated with either Endo-Porter, MO, or both. Three biological replicates of each treatment were performed.

2.3. Cell Counts and Fluorescence Quantification

Following the addition of 4 μM Endo-Porter and 1 μM or 10 μM MO, the viability of cultures of *A. carterae* were observed over a 96-h period by measuring their autofluorescence via flow cytometry. Measurements of the cultures within the first hour of treatment were labelled as Hour 0. Cell counts for each condition were determined on a BD C6 Accuri Flow Cytometer (BD Biosciences, San Jose, CA, USA), equipped with laser excitation at 488 and 640 nm and emission at 533/30, 585/40, and >670 nm. The FSC-A and fluorescence channels were used to select for live cells; from this selection, cells with Lissamine emission signals were detected (585 nm). Cells were grouped into low and high Lissamine fluorescence, with high fluorescence showing an intensity minimum of 10^4 relative fluorescence units (RFUs).

2.4. Cell Imaging

Images of A. carterae cells with and without MO treatment were taken on a STELLARIS confocal microscope (Leica Microsystems, South San Francisco, CA, USA), equipped with 405, 552 and 638 nm lasers, and PMT and HyD detectors collecting emission within 590–600 nm and 680–720 nm, respectively.

2.5. Quantification of Protein Expression

Protein expression was quantified by Western blotting. For the initial Western blot analyses done for the 1 μM concentration of morpholinos, the cell density of the cultures were quantified via flow-cytometry to create equal volume pellets containing ~75,000 cells for each condition 48 h post-treatment. A NuPAGE 4 to 12% Bis-Tris 1.5 mm Mini Protein Gels was used. Cell pellets were prepared for electrophoresis in 3× sample buffer (Blue Loading Buffer Pack, New England Biolabs, Ipswich, MA, USA), heated to 96 °C for 10 min and centrifuged for 2 min at 10,000× *g*; from the 15 μL total, 10 μL of each extract was electrophoresed at 165 V until the dye front reached the bottom of the gel. The gel was transferred to a membrane with the Trans Blot Turbo Transfer system. Protein loading and relative expression levels were verified by probing the same blot with anti-eIF4E-1a mouse monoclonal and HRP-conjugated anti-mouse IgG. The labeled bands, as well as band intensities, were detected with ImageLab software.

For the subsequent experiment testing the higher 10 μM concentration of morpholinos, equal quantities of whole-cell lysates containing 100,000 cells were prepared from each of the triplicate sample cultures at 48 and 96 h. Samples were once again split equally over two NuPAGE 4–12% Bis-Tris Gels and run for 50 min at 165 V. Both gels were transferred to a membrane with the Trans Blot Turbo Transfer system. To control for cell loading error, total protein was detected on one membrane using No-Stain Protein labelling Reagent (Invitrogen, Waltham, MA, USA). Summation of total protein in each lane was found using ImageJ. Protein loading and relative expression levels were again verified by probing the second blot with anti-eIF4E-1a mouse monoclonal and HRP-conjugated anti-mouse IgG. The labeled bands were detected with ImageLab software, as well as band intensities. The relative production of eIF4E-1a was analyzed by dividing the eIF4E-1a volume by the summation of total protein.

2.6. Statistical Considerations

All conditions were performed in triplicate. Pairwise sample comparisons within timepoints were analyzed in R-Studio with *t*-tests using pooled standard deviations [52]. A *p*-value of <0.05 was considered statistically significant.

3. Results

3.1. Cell Viability

Immediately post-treatment, the cultures displayed a suspension of growth, but recovery was observed at 24 h (Figure 1). The cultures containing the antisense morpholino also showed significant differences in growth compared to the control immediately after treatment addition and at both 24 and 48 h.

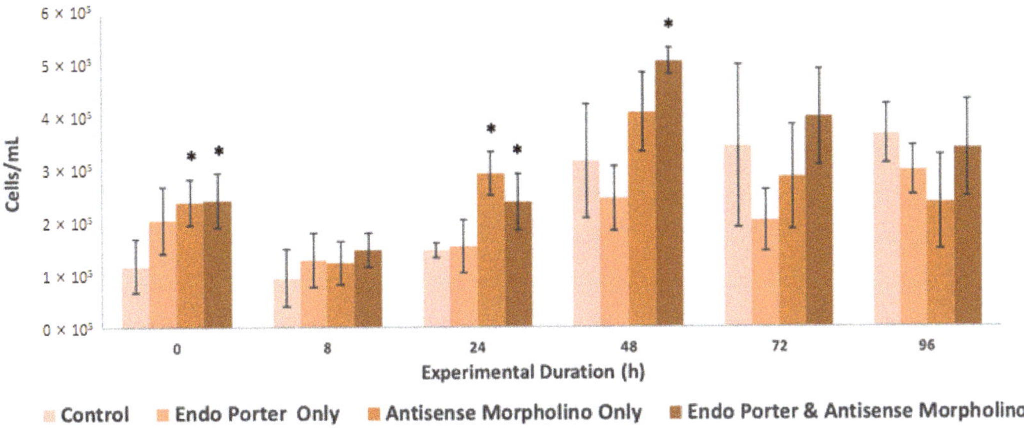

Figure 1. Flow cytometer population counts from *Amphidinium carterae* cultures (N = 3). Conditions included the control, Endo-Porter only, antisense morpholino only, and the Endo-Porter and antisense morpholino combined. Statistical analysis was done using *t*-tests with pooled standard deviations. * Significantly different from 'Control' based on a *p*-value of < 0.05.

3.2. Uptake of the Morpholino

The intensity of the Lissamine signal within cells was measured by flow cytometry (Figure 2). Peak Lissamine fluorescence was observed at 48 h, with about 13% of the population uptaking a high amount of fluorescently-tagged morpholinos [47], and waned after this timepoint. The mean Lissamine fluorescence per cell with the Endo-Porter delivery system plus the MO was over 50X greater than that of the control, and the median was over 1.5X greater, showing significant uptake of the MO in *A. carterae*, as well as a large positively-skewed distribution of uptake efficiency (Supplementary Table S1). Cells were also able to uptake morpholino without Endo Porter, but to a significantly lower degree, with no correlation to experimental duration.

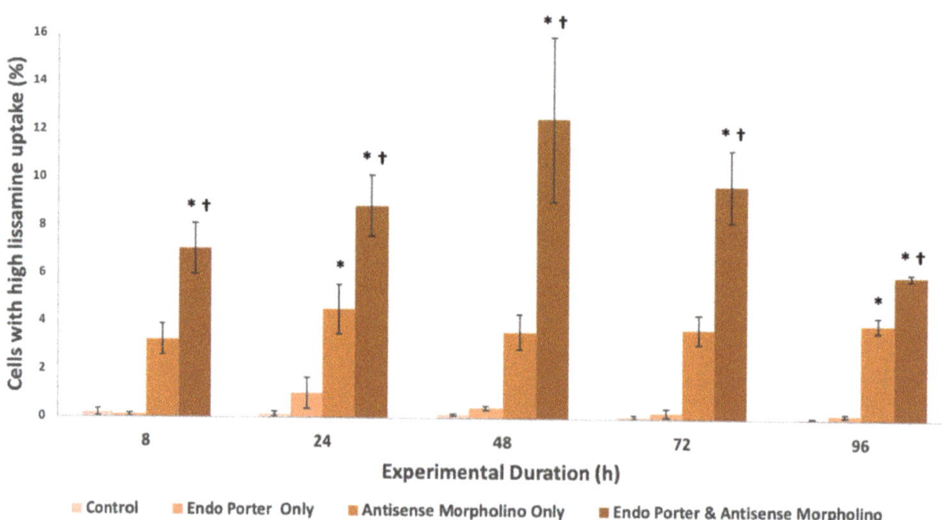

Figure 2. Percent of A.carterae populations with high Lissamine uptake after 48 h (cutoff 104 RFUs). Conditions included the control, Endo Porter only, antisense morpholino only, and the Endo Porter and antisense morpholino combined (N = 3). Statistical analysis was done using *t*-tests with pooled standard deviations. * Significantly different from 'Control' based on a *p*-value of < 0.05. † Significant difference between "Antisense Morpholino Only" and "Endo Porter & Anti-sense Morpholino" based on a *p*-value of < 0.05.

Images of cells were also captured by confocal microscopy (Figure 3). Lissamine fluorescence was detected within the range of 590–600 nm (peak 593 nm), and autofluorescence was measured between 700–720 nm. Confocal images display a diffuse pseudo-blue coloring within the cytoplasm of the dinoflagellate cells, as well as in a large area near the nucleus, indicating both diffuse and localized morpholino presence (Figure 3).

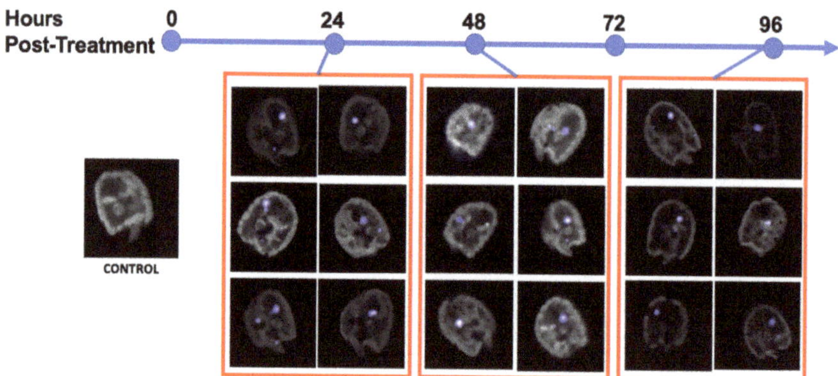

Figure 3. Confocal images of *Amphidinium carterae*. A control cell with no antisense morpholino introduced is on the far left. Red boxes encompass images of cells from a culture with Endo Porter and antisense-morpholino, fluorescently tagged with Lissamine (593 nm, pseudo-blue) at 24-, 48- and 96-h post-treatment. The grey areas are the auto-fluorescence emitted within 720–759 nm.

3.3. Initial Western Blot Analysis

From the initial Western Blot analysis, we found a statistically significant decrease in the expression of eIF4E-1a of about 30% compared to the control in cultures containing both the 1 µM morpholino and Endo Porter (Figure 4). Cultures with Endo Porter showed a decrease of 11% in eIF4E-1a expression over cultures without Endo Porter, although the difference was not statistically significant.

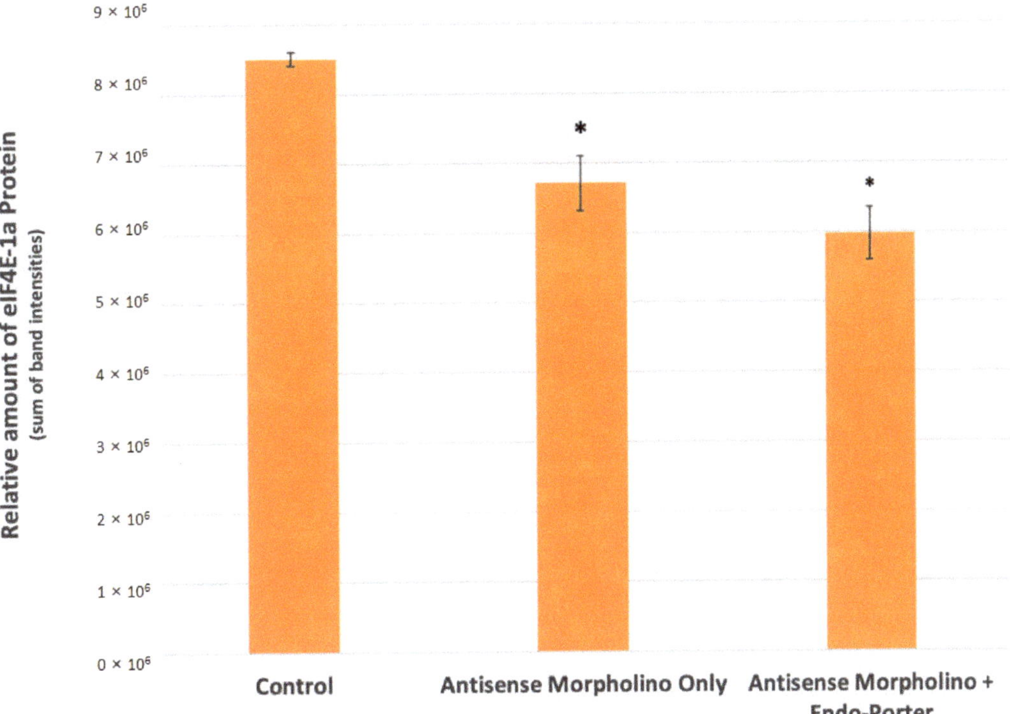

Figure 4. Western blot analyses for eIF4E-1a concentrations within control and treated cells at 48 h post-MO [1 µM] addition (N = 3). Protein loading and relative expression levels was verified by probing with anti-eIF4E-1a mouse monoclonal and HRP-conjugated an-ti-mouse IgG. eIF4E-1a Protein area volumes were reduced by 30% after introduction of custom translation blocking morpholino and Endo Porter after 48 h. Statistical analy-sis was done using *t*-tests with pooled standard deviations. * Significantly different from "Control" based on a *p*-value of < 0.05. No significant difference found between "Antisense Morpholino Only" and "Antisense Morpholino + Endo Porter" based on a *p*-value of < 0.05.

3.4. Increase in MO Concentration

Once the concentration of morpholinos was increased from 1 µM to 10 µM, we found that with the custom eIF4E-1a target morpholino and Endo Porter there was a decrease in population eIF4E-1a expression at 48 h compared to the control of about 42% (Figure 5). This was an increase from the 30% found when using only 1 µM of MO, constituting a total 11.8% decrease with the increased MO concentration (Table 1). The cultures containing eIF4E-1a-target morpholino only also showed reduced target-protein production compared to the control of 21%. When compared to the culture containing only the standard non-target morpholino and Endo Porter, both of the cultures containing the eIF4E-1a-target morpholino with and without the Endo Porter appeared to exhibit a significant decrease of 47% and 29%, respectively (Figure 5). No significant difference was found between the cultures containing eIF4E-1a-target morpholino with or without the Endo Porter, although the cultures with the Endo Porter showed an average decrease of 25% compared to the cultures without.

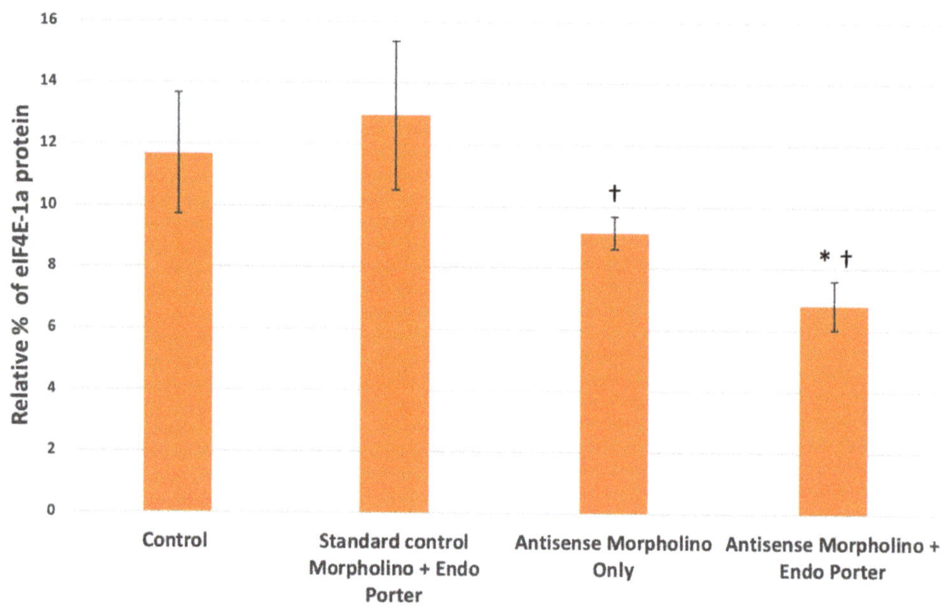

Figure 5. Western blot analyses for eIF4E-1a concentrations within control and treated cells at 48 post-MO [10 μM] addition (N = 3). Protein loading and relative expression levels was veri-fied by probing with anti-eIF4E-1a mouse monoclonal and HRP-conjugated anti-mouse IgG and compared to total protein volumes. Relative eIF4E-1a levels are lower in Am-phidinium population after 48 h of being subjected to custom translation-blocking morpholinos. Statistical analysis was done using *t*-tests with pooled standard deviations. * Significantly different from 'Control' based on a *p*-value of < 0.05. † Significantly different from "Standard Morpholino + Endo Porter" based on a *p*-value of < 0.05. No significant difference found between "Antisense Morpholino Only" and "Antisense Morpholino + Endo Porter" based on a *p*-value of < 0.05.

Table 1. Volume of eIF4E-1a produced compared to the control. Volumes of eIF4E-1a were quantified via western blot and compared to the density of a whole protein stain per sample (N = 3). Percent of eIF4E-1a for each sample were compared to the control sample to describe relative eIF4E-1a production. Statistical analysis was done using *t*-tests with pooled standard deviations.

Morpholino Concentration	1 μM	10 μM	
	Relative eIF4E-1a Production		**Percent Change**
Antisense Morpholino Only	78.8% *	78.3%	0.5%
Antisense Morpholino + Endo Porter	70.1% *	58.3% *	11.8%

* Significant difference when compared to the control based on a *p*-value of < 0.05.

Interestingly, we also found that expression levels became relatively similar after 96 h, showing the temporary effects of the morpholino (Supplemental Figure S1).

4. Discussion

Here we have shown that novel delivery peptide technology has allowed for successful introduction of custom antisense morpholinos into *A. carterae* cells without significantly decreasing viability. This method delivers substances into the cytosol of cells by an endocytosis-mediated process that avoids damaging the plasma membrane of the cell [47,53]. With this new system, exploration into the key players of different metabolic pathways may be closer than expected for dinoflagellates.

The above data was collected using the manufacturer recommended concentrations of reagents, but as we look further into these systems, we hope to optimize the outcome by adjusting protocol values. Currently, with 1 μM of MO added, we have been able to successfully introduce a high concentration of the MO into approximately 13% of the population (Figure 2), with a significant decrease in eIF4E-1a protein production. Dose-dependent effects of the MOs have also been observed, with an increase in MO concentration from 1 μM to 10 μM resulting in a decrease in eIF4E-1a from 30% to 42%; an 11.83% decrease in total (Table 1). Optimistically we would like to increase this percentage to a level where protein production is functionally suspended, or be able to separate cells from the population based on their MO uptake.

Interestingly. western blot analyses for both the initial 1 μM and subsequent 10 μM MO experiment showed no statistically significant difference between the cultures with and without Endo-Porter, although the cultures with the Endo-Porter showed consistently lower average eIF4E-1a protein produced (Figures 4 and 5). This would indicate that even without a delivery peptide, *A. carterae* cells are able to uptake the MO. Reasons for MO uptake without a delivery peptide are still unclear. Over a million identified peptide sequences within various dinoflagellates are still of unknown function and origin, making their evolutionary history ambiguous [54]. Dinoflagellates are thought to have undergone multiple organellogensis events, where the genome of endosymbiotic algae becomes a plastid and/or genes from the endosymbiont are transferred to the nucleus [55]. The evolutionary nature of dinoflagellates to accept foreign genes appears to be high [56]. One theory for why the *A. carterae* cells took up the unaided MO could be the possibility that dinoflagellate systems are more open to horizontal gene transfer (HGT) than previously imagined [57]. Recent studies have shown that genomes within dinoflagellates may be more open to foreign contributions from both bacteria and eukaryotes compared to other organisms [57,58]. Further research needs to be conducted to follow how foreign genetic material is received by dinoflagellates, as well as how transcripts are processed.

A wide distribution of MO uptake may account for the discrepancy of measured eIF4E-1a protein production compared to the cellular MO uptake (Supplemental Table S1). Although we suggest about 13% of the cells have a high uptake after addition of 1 μM of MO, the eIF4E-1a production was decreased by about 30% according to the western results (Figure 4). We assume that there is a range of efficiency uptake within the population, so some cells may be producing little or no eIF4E-1a, while others may be producing their average amounts. Once again, changes to the concentrations used or the use of a cell-sorter may be necessary to observe higher gene knockdown efficiency.

MO localization does occur within the *A. carterae* cells, usually in a large area by the nucleus (Figure 3). In previous studies, Endo-Porter sometimes results in unsuccessful endosomal acidification, and therefore no release of the MO into the cytoplasm from their vesicle, but these localization points are usually known to appear as punctate fluorescence throughout the cell [47]. The singular, large area of localization may signify MO aggregation in the nucleolus or an RNA-granule [55,59–62]. Dinoflagellates are known to have very unusual nuclei, specially named the "dinokarya." Among other peculiar features are recently discovered "nuclear tunnels" which extend from the nuclear envelope of *Polykrikos kofoidii*, specifically during mitosis [59]. These nuclear envelope tunnels are also connected with a membranous structure throughout the nucleus known as the "nuclear net". The discovery of these structures adds a new level of complexity to the dinoflagellate nuclear membrane, and may allude to more complex processing for transcription and translation.

The unusual nuclear envelope tunnels along with our MO localization could very well be connected. Since dinoflagellates are known to regulate gene expression at the translational level, this cellular organization of mRNA may be a crucial step for gene expression, which needs to be analyzed further.

In theory, this method of gene knockdown is progress towards expanding research into the biosynthetic pathways within dinoflagellate. Currently, this research is sorely lacking, largely due to their complex genomes and unusual cell biology. Translation regulation is now understood as a crucial step in gene expression, far beyond that of transcriptional control [22,48]. In addition, further understanding of the components of the translation machinery is required to understand the expression of specific genes, which make up the dinoflagellate "translational toolkit". As we work to optimize this procedure, other pathways could be targeted as well in other species of dinoflagellates.

One interesting area would be to target the transcript responsible for toxin production. The function of toxins produced by dinoflagellates has been theorized but is still unknown. Theories include allelopathy, prey-capture, and as a defense [18]. One way to discern the function of the toxins would be to knockdown their expression to see how this affects feeding and swimming behavior. Mixotrophic characteristics of dinoflagellate species known to produce toxins could be monitored through photosynthesis and respiration rates, as well as swimming behavior via digital holographic microscopy [63]. This data could produce more evidence to the intended functionality of dinoflagellate toxins.

5. Conclusions

This primary study provides proof-of-principle for the possibility to specifically down-regulate gene expression in dinoflagellates using antisense morpholinos and a novel delivery system. Additional work is necessary to validate and optimize these findings, and to extend them to other biosynthetic pathways. Further work will also be required to investigate the effects of gene knockdown on the various eIF4E family members in order to possibly unveil a translation toolkit used by dinoflagellates to regulate gene expression. Studies involving toxin biosynthesis pathways would also benefit from successful knockdowns in order to identify functionality and key synthesis steps.

Supplementary Materials: The following supporting information can be downloaded at: https://www.mdpi.com/article/10.3390/microorganisms10061131/s1, Figure S1: Western blot analyses for eIF4E-1a concentrations within control and treated cells at 96 h post-MO [10 μM] addition; Table S1: Lissamine uptake in *A. carterae* cells 48 h post treatment.

Author Contributions: Conceptualization, review and editing by M.J. and A.R.P.; Methodology by M.J. and A.R.P.; Supervision by A.R.P.; Formal analysis and writing by M.J. All authors have read and agreed to the published version of the manuscript.

Funding: This research received no external funding.

Institutional Review Board Statement: Not applicable.

Informed Consent Statement: Not applicable.

Data Availability Statement: Not applicable.

Acknowledgments: I would like to thank the Ernest Williams for his wealth of knowledge and advice, as well as helping me finalize this project. This is contribution #6181 from UMCES and #22-104 from IMET.

Conflicts of Interest: The authors declare no conflict of interest.

References

1. Not, F.; Siano, R.; Kooistra, W.H.C.F.; Simon, N.; Vaulot, D.; Probert, I. Diversity and Ecology of Eukaryotic Marine Phyto-plankton. In *Advances in Botanical Research*; Elsevier: Amsterdam, The Netherlands, 2012.
2. Adolf, J.E.; Stoecker, D.K.; Harding, L.W. The balance of autotrophy and heterotrophy during mixotrophic growth of Kar-lodinium micrum (*Dinophyceae*). *J. Plankton Res.* **2006**, *28*, 737–751. [CrossRef]

3. Place, A.R.; Bowers, H.A.; Bachvaroff, T.R.; Adolf, J.E.; Deeds, J.R.; Sheng, J. Karlodinium veneficum-The little dinoflagellate with a big bite. *Harmful Algae* **2012**, *14*, 179–195. [CrossRef]

4. Botana, L.M. *Seafood and Freshwater Toxins: Pharmacology, Physiology, and Detection*, 2nd ed.; CRC Press: Boca Raton, FL, USA, 2008; pp. 1–943.

5. Cusick, K.D.; Widder, E.A. Intensity differences in bioluminescent dinoflagellates impact foraging efficiency in a nocturnal predator. *Bull. Mar. Sci.* **2014**, *90*, 797–811. [CrossRef]

6. Decelle, J.; Carradec, Q.; Pochon, X.; Henry, N.; Romac, S.; Mahé, F.; Dunthorn, M.; Kourlaiev, A.; Voolstra, C.R.; Wincker, P.; et al. Worldwide Occurrence and Activity of the Reef-Building Coral Symbiont Symbiodinium in the Open Ocean. *Curr. Biol.* **2018**, *28*, 3625–3633.e3. [CrossRef] [PubMed]

7. Aquino-Cruz, A.; Okolodkov, Y.B. Impact of increasing water temperature on growth, photosynthetic efficiency, nutrient consumption, and potential toxicity of Amphidinium cf. carterae and Coolia monotis (Dinoflagellata). *Rev. Biol. Mar. Oceanogr.* **2016**, *51*, 565–580. [CrossRef]

8. Gobler, C.J. Climate Change and Harmful Algal Blooms: Insights and perspective. *Harmful Algae* **2020**, *91*, 101731. [CrossRef] [PubMed]

9. Griffith, A.W.; Gobler, C.J. Harmful algal blooms: A climate change co-stressor in marine and freshwater ecosystems. *Harmful Algae* **2020**, *91*, 101590. [CrossRef] [PubMed]

10. Verma, A.; Barua, A.; Ruvindy, R.; Savela, H.; Ajani, P.A.; Murray, S.A. The genetic basis of toxin biosynthesis in dinoflagel-lates. *Microorganisms* **2019**, *7*, 222. [CrossRef]

11. Wells, M.L.; Trainer, V.L.; Smayda, T.J.; Karlson, B.S.; Trick, C.G.; Kudela, R.M.; Ishikawa, A.; Bernard, S.; Wulff, A.; Ander-son, D.M.; et al. Harmful algal blooms and climate change: Learning from the past and present to forecast the future. *Harmful Algae* **2015**, *49*, 68–93. [CrossRef] [PubMed]

12. Pagliara, P.; Caroppo, C. Toxicity assessment of *Amphidinium carterae, Coolia* cfr. *monotis* and *Ostreopsis* cfr. *ovata* (Di-nophyta) isolated from the northern Ionian Sea (Mediterranean Sea). *Toxicon* **2012**, *60*, 1203–1214. [CrossRef] [PubMed]

13. Brownlee, E.F.; Sellner, S.G.; Sellner, K.G.; Nonogaki, H.; Adolf, J.E.; Bachvaroff, T.R.; Place, A.R. Responses of *Crassostrea virginica* (Gmelin) and *C. ariakensis* (Fujita) To Bloom-Forming Phytoplankton Including Ichthyotoxic *Karlodinium veneficum* (Ballantine). *J. Shellfish. Res.* **2008**, *27*, 581–591. [CrossRef]

14. Galimany, E.; Place, A.R.; Ramón, M.; Jutson, M.; Pipe, R.K. The effects of feeding Karlodinium veneficum (PLY # 103; Gymno-dinium veneficum Ballantine) to the blue mussel Mytilus edulis. *Harmful Algae* **2008**, *7*, 91–98.

15. Baden, D.G. Brevetoxins: Unique polyether dinoflagellate toxins. *FASEB J.* **1989**, *3*, 1807–1817. [CrossRef] [PubMed]

16. Deeds, J.R.; Hoesch, R.E.; Place, A.R.; Kao, J.P. The cytotoxic mechanism of karlotoxin 2 (KmT $\times$ 2) from Karlodinium veneficum (Dinophyceae). *Aquat. Toxicol.* **2015**, *159*, 148–155. [CrossRef] [PubMed]

17. Peng, J.; Place, A.R.; Yoshida, W.; Anklin, C.; Hamann, M.T. Structure and absolute configuration of karlotoxin-2, an ichthy-otoxin from the marine dinoflagellate karlodinium veneficum. *J. Am. Chem. Soc.* **2010**, *132*, 3277–3279. [CrossRef]

18. Sheng, J.; Malkiel, E.; Katz, J.; Adolf, J.E.; Place, A.R. A dinoflagellate exploits toxins to immobilize prey prior to ingestion. *Proc. Natl. Acad. Sci. USA* **2010**, *107*, 2082–2087. [CrossRef]

19. Van Wagoner, R.M.; Deeds, J.R.; Satake, M.; Ribeiro, A.A.; Place, A.R.; Wright, J.L. Isolation and characterization of karlo-toxin 1, a new amphipathic toxin from Karlodinium veneficum. *Tetrahedron Lett.* **2008**, *49*, 6457–6461. [CrossRef] [PubMed]

20. Walsh, P.J. *Oceans and Human Health: Risks and Remedies from the Seas*; Academic Press: Cambridge, MA, USA, 2008.

21. Hieda, M.; Sorada, A.; Kinoshita, M.; Matsumori, N. Amphidinol 3 preferentially binds to cholesterol in disordered do-mains and disrupts membrane phase separation. *Biochem. Biophys. Rep.* **2021**, *26*, 100941. [PubMed]

22. Jones, G.D.; Williams, E.P.; Place, A.R.; Jagus, R.; Bachvaroff, T.R. The alveolate translation initiation factor 4E family reveals a custom toolkit for translational control in core dinoflagellates. *BMC Evol. Biol.* **2015**, *15*, 14. [CrossRef] [PubMed]

23. Brayton, K.A.; Lau, A.O.; Herndon, D.R.; Hannick, L.; Kappmeyer, L.S.; Berens, S.J.; Bidwell, S.L.; Brown, W.C.; Crabtree, J.; Fadrosh, D.; et al. Genome sequence of Babesia bovis and comparative analysis of apicomplexan hemoprotozoa. *PLoS Pathog.* **2007**, *3*, 1401–1413. [CrossRef] [PubMed]

24. Ebenezer, T.E.; Zoltner, M.; Burrell, A.; Nenarokova, A.; Vanclová, A.M.G.N.; Prasad, B.; Soukal, P.; Santana-Molina, C.; O'Neill, E.; Nankissoor, N.N.; et al. Transcriptome, proteome and draft genome of Euglena gracilis. *BMC Biol.* **2019**, *17*, 11. [CrossRef] [PubMed]

25. Zhang, C.; Lin, S. Initial evidence of functional siRNA machinery in dinoflagellates. *Harmful Algae* **2019**, *81*, 53–58. [CrossRef] [PubMed]

26. Amorim, I.S.; Lach, G.; Gkogkas, C.G. The Role of the Eukaryotic Translation Initiation Factor 4E (eIF4E) in Neuropsychiat-ric Disorders. *Front. Genet.* **2018**, *9*, 561. [CrossRef] [PubMed]

27. Batool, A.; Aashaq, S.; Andrabi, K.I. Eukaryotic initiation factor 4E (eIF4E): A recap of the cap-binding protein. *J. Cell. Biochem.* **2019**, *120*, 14201–14212. [CrossRef] [PubMed]

28. Altmann, M.; Müller, P.P.; Pelletier, J.; Sonenberg, N.; Trachsel, H. A Mammalian Translation Initiation Factor Can Substi-tute for Its Yeast Homologue in Vivo. *J. Biol. Chem.* **1989**, *264*, 12145–12147. [CrossRef]

29. Feoktistova, K.; Tuvshintogs, E.; Do, A.; Fraser, C.S. Human eIF4E promotes mRNA restructuring by stimulating eIF4A hel-icase activity. *Proc. Natl. Acad. Sci. USA* **2013**, *110*, 13339–13344. [CrossRef]

30. Ibrahimo, S.; Holmes, L.E.; Ashe, M.P. Regulation of translation initiation by the yeast eIF4E binding proteins is required for the pseudohyphal response. *Yeast* **2006**, *23*, 1075–1088. [CrossRef]
31. Joshi, B.; Lee, K.; Maeder, D.L.; Jagus, R. Phylogenetic analysis of eIF4E-family members. *BMC Evol. Biol.* **2005**, *5*, 48. [CrossRef]
32. Piserà, A.; Campo, A.; Campo, S. Structure and functions of the translation initiation factor eIF4E and its role in cancer development and treatment. *J. Genet. Genom.* **2018**, *45*, 13–24. [CrossRef]
33. Grifo, J.A.; Tahara, S.M.; Leis, J.P.; Morgan, M.A.; Shatkin, A.J.; Merrick, W.C. Characterization of eukaryotic initiation factor 4A, a protein involved in ATP-dependent binding of globin mRNA. *J. Biol. Chem.* **1982**, *257*, 5246–5252. [CrossRef]
34. Hernández, G.; Vazquez-Pianzola, P. Functional diversity of the eukaryotic translation initiation factors belonging to eIF4 families. *Mech. Dev.* **2005**, *122*, 865–876. [CrossRef] [PubMed]
35. Joshi, B.; Cameron, A.; Jagus, R. Characterization of mammalian eIF4E-family members. *Eur. J. Biochem.* **2004**, *271*, 2189–2203. [CrossRef] [PubMed]
36. Lasko, P. The drosophila melanogaster genome: Translation factors and RNA binding proteins. *J. Cell Biol.* **2000**, *150*, F51–F56. [CrossRef] [PubMed]
37. Bachvaroff, T.R.; Place, A.R. From stop to start: Tandem gene arrangement, copy number and Trans-splicing sites in the di-noflagellate *Amphidinium carterae*. *PLoS ONE* **2008**, *3*, e2929. [CrossRef] [PubMed]
38. Beauchemin, M.; Roy, S.; Daoust, P.; Dagenais-Bellefeuille, S.; Bertomeu, T.; Letourneau, L.; Lang, B.F.; Morse, D. Dinoflag-ellate tandem array gene transcripts are highly conserved and not polycistronic. *Proc. Natl. Acad. Sci. USA* **2012**, *109*, 15793–15798. [CrossRef]
39. Li, L.; Wang, C.C. Identification in the ancient protist Giardia lamblia of two eukaryotic translation initiation factor 4E homologues with distinctive functions. *Eukaryot. Cell* **2005**, *4*, 948–959. [CrossRef]
40. Jagus, R.; Bachvaroff, T.R.; Joshi, B.; Place, A.R. Diversity of eukaryotic translational initiation factor eIF4E in protists. *Comp. Funct. Genom.* **2012**, *2012*, 134839. [CrossRef]
41. Morf, L.; Pearson, R.J.; Wang, A.S.; Singh, U. Robust gene silencing mediated by antisense small RNAs in the pathogenic protist Entamoeba histolytica. *Nucleic Acids Res.* **2013**, *41*, 9424–9437. [CrossRef]
42. Kwok, A.C.M.; Mak, C.C.M.; Wong, F.T.W.; Wong, J.T.Y. Novel method for preparing spheroplasts from cells with an in-ternal cellulosic cell wall. *Eukaryot. Cell* **2007**, *6*, 563–567. [CrossRef]
43. Yan, T.H.K.; Wu, Z.; Kwok, A.C.M.; Wong, J.T.Y. Knockdown of dinoflagellate condensin CcSMC4 subunit leads to s-phase impediment and decompaction of liquid crystalline chromosomes. *Microorganisms* **2020**, *8*, 565. [CrossRef]
44. Chan, W.S.; Kwok, A.C.M.; Wong, J.T.Y. Knockdown of dinoflagellate cellulose synthase CesA1 resulted in malformed in-tracellular cellulosic thecal plates and severely impeded cyst-to-swarmer transition. *Front. Microbiol.* **2019**, *10*, 546. [CrossRef] [PubMed]
45. Akbar, M.A.; Ahmad, A.; Usup, G.; Bunawan, H. Current knowledge and recent advances in marine dinoflagellate tran-scriptomic research. *J. Mar. Sci. Eng.* **2018**, *6*, 13. [CrossRef]
46. Baumgarten, S.; Bayer, T.; Aranda, M.; Liew, Y.J.; Carr, A.; Micklem, G.; Voolstra, C.R. Integrating microRNA and mRNA expression profiling in *Symbiodinium microadriaticum*, a dinoflagellate symbiont of reef-building corals. *BMC Genom.* **2013**, *14*, 704. [CrossRef] [PubMed]
47. Summerton, J.E. Endo-Porter: A Novel Reagent for Safe, Effective Delivery of Substances into Cells. *Ann. N. Y. Acad. Sci.* **2005**, *1058*, 62–75. [CrossRef] [PubMed]
48. Roy, S.; Jagus, R.; Morse, D. Translation and translational control in dinoflagellates. *Microorganisms* **2018**, *6*, 30. [CrossRef] [PubMed]
49. Berges, J.A.; Franklin, D.J.; Harrison, P.J. Evolution of an artificial seawater medium: Improvements in enriched seawater, artificial water over the last two decades. *J. Phycol.* **2001**, *37*, 1138–1145. [CrossRef]
50. Liu, C.L.; Place, A.R.; Jagus, R. Use of antibiotics for maintenance of axenic cultures of *Amphidinium carterae* for the analy-sis of translation. *Mar. Drugs* **2017**, *15*, 242. [CrossRef]
51. Todaro, A.M.; Hackeng, T.M.; Castoldi, E. Antisense-mediated down-regulation of factor v-short splicing in a liver cell line model. *Appl. Sci.* **2021**, *11*, 9621. [CrossRef]
52. The R Development Core Team. R: A Language and Environment for Statistical Computing. 2013. Available online: http://www.R-Project.org/ (accessed on 24 May 2022).
53. Mellert, K.; Lamla, M.; Scheffzek, K.; Wittig, R.; Kaufmann, D. Enhancing Endosomal Escape of Transduced Proteins by Photochemical Internalisation. *PLoS ONE* **2012**, *7*. [CrossRef]
54. Stephens, T.G.; Ragan, M.A.; Bhattacharya, D.; Chan, C.X. Core genes in diverse dinoflagellate lineages include a wealth of conserved dark genes with unknown functions. *Sci. Rep.* **2018**, *8*, 17175. [CrossRef]
55. Sarai, C.; Tanifuji, G.; Nakayama, T.; Kamikawa, R.; Takahashi, K.; Yazaki, E.; Matsuo, E.; Miyashita, H.; Ishida, K.-I.; Iwataki, M.; et al. Dinoflagellates with relic endosymbiont nuclei as models for elucidating organellogenesis. *Proc. Natl. Acad. Sci. USA* **2020**, *117*, 5364–5375. [CrossRef] [PubMed]
56. Morden, C.W.; Sherwood, A.R. Continued evolutionary surprises among dinoflagellates. *Proc. Natl. Acad. Sci. USA* **2002**, *99*, 11558–11560. [CrossRef] [PubMed]
57. Mackiewicz, P.; Bodyƚ, A.; Moszczyński, K. The case of horizontal gene transfer from bacteria to the peculiar dinoflagellate plastid genome. *Mob. Genet. Elem.* **2013**, *3*, e25845. [CrossRef] [PubMed]

58.   Wisecaver, J.H.; Brosnahan, M.L.; Hackett, J.D. Horizontal gene transfer is a significant driver of gene innovation in dino-flagellates. *Genome Biol. Evol.* **2013**, *5*, 2368–2381. [CrossRef] [PubMed]

59.   Gavelis, G.S.; Herranz, M.; Wakeman, K.C.; Ripken, C.; Mitarai, S.; Gile, G.H.; Keeling, P.J.; Leander, B.S. Dinoflagellate nu-cleus contains an extensive endomembrane network, the nuclear net. *Sci. Rep.* **2019**, *9*, 839. [CrossRef] [PubMed]

60.   Gornik, S.G.; Hu, I.; Lassadi, I.; Waller, R.F. The biochemistry and evolution of the dinoflagellate nucleus. *Microorganisms* **2019**, *7*, 245. [CrossRef]

61.   Lee, C.Y.S.; Putnam, A.; Lu, T.; He, S.; Ouyang, J.P.T.; Seydoux, G. Recruitment of mRNAs to P granules by condensation with intrinsically-disordered proteins. *eLife* **2020**, *9*, e52896. [CrossRef]

62.   Soyer-Gobillard, M.O.; Geraud, M.L. Nucleolus behaviour during the cell cycle of a primitive dinoflagellate eukaryote, Prorocen-trum micans Ehr., seen by light microscopy and electron microscopy. *J. Cell Sci.* **1992**, *102*, 475–485. [CrossRef]

63.   Sheng, J.; Malkiel, E.; Katz, J.; Adolf, J.; Belas, R.; Place, A.R. Digital holographic microscopy reveals prey-induced changes in swimming behavior of predatory dinoflagellates. *Proc. Natl. Acad. Sci. USA* **2007**, *104*, 17512–17517. [CrossRef]

 *microorganisms*

Article

Dinoflagellate Phosphopantetheinyl Transferase (PPTase) and Thiolation Domain Interactions Characterized Using a Modified Indigoidine Synthesizing Reporter

Ernest Williams *, Tsvetan Bachvaroff and Allen Place 

Institute of Marine and Environmental Technologies, University of Maryland Center for Environmental Science, 701 East Pratt St., Baltimore, MD 21202, USA; bachvarofft@umces.edu (T.B.); place@umces.edu (A.P.)
* Correspondence: williamse@umces.edu; Tel.: +1-410-234-8829

Abstract: Photosynthetic dinoflagellates synthesize many toxic but also potential therapeutic compounds therapeutics via polyketide/non-ribosomal peptide synthesis, a common means of producing natural products in bacteria and fungi. Although canonical genes are identifiable in dinoflagellate transcriptomes, the biosynthetic pathways are obfuscated by high copy numbers and fractured synteny. This study focuses on the carrier domains that scaffold natural product synthesis (thiolation domains) and the phosphopantetheinyl transferases (PPTases) that thiolate these carriers. We replaced the thiolation domain of the indigoidine producing BpsA gene from *Streptomyces lavendulae* with those of three multidomain dinoflagellate transcripts and coexpressed these constructs with each of three dinoflagellate PPTases looking for specific pairings that would identify distinct pathways. Surprisingly, all three PPTases were able to activate all the thiolation domains from one transcript, although with differing levels of indigoidine produced, demonstrating an unusual lack of specificity. Unfortunately, constructs with the remaining thiolation domains produced almost no indigoidine and the thiolation domain for lipid synthesis could not be expressed in *E. coli*. These results combined with inconsistent protein expression for different PPTase/thiolation domain pairings present technical hurdles for future work. Despite these challenges, expression of catalytically active dinoflagellate proteins in *E. coli* is a novel and useful tool going forward.

Keywords: dinoflagellate; PKS; phosphopantetheinyl transferase; toxin; BpsA; indigoidine; natural product

Citation: Williams, E.; Bachvaroff, T.; Place, A. Dinoflagellate Phosphopantetheinyl Transferase (PPTase) and Thiolation Domain Interactions Characterized Using a Modified Indigoidine Synthesizing Reporter. *Microorganisms* **2022**, *10*, 687. https://doi.org/10.3390/microorganisms10040687

Academic Editor: Shauna Murray

Received: 20 January 2022
Accepted: 14 March 2022
Published: 23 March 2022

Publisher's Note: MDPI stays neutral with regard to jurisdictional claims in published maps and institutional affiliations.

1. Introduction

Dinoflagellates make a variety of natural products that have largely been identified based on their impact to human and animal health [1–4]. The actual biological and/or ecological roles are largely unknown and require further study. The exceptions include karlotoxin, the only toxin known to be actively released from the cell for prey capture and predator avoidance [5,6], and brevetoxin that likely functions as an indicator of redox state in the chloroplast [7,8]. This functional knowledge gap is exacerbated by a lack of a biosynthetic framework that would allow a more thorough cataloging of the natural products produced by dinoflagellates as well as insights into their evolution.

Natural product synthesis has been extensively studied in bacteria and fungi, yielding a mechanistic framework that operates as a series of modules with repeated chemistries followed by some modifications resulting in the final molecule. Essentially, small carboxylic acids are added to the thiol end of a phosphopantetheinate group attached to the serine of a carrier protein [9] via a condensation reaction that releases either carbon dioxide or water with prior activation by ATP [10–12]. These building blocks are then modified by subsequent reduction, methylation, carbon deletion, and other rarer reactions before the next carboxylic acid is added. In general, these are added by genetic modules comprised of single proteins with multiple functional domains or multiple cis-acting proteins brought together to form an enzymatic complex, although trans-acting elements are not uncommon [13–15] and substrates from multiple pathways can be combined [16,17].

Research into the biosynthesis of many natural products has relied heavily on the fact that gene arrangement is strongly predictive of a given natural products' final structure. Unfortunately, dinoflagellate genomes are large and heavily duplicated [18], although mass spectrometry and NMR have been able to readily identify that dinoflagellate toxins have the hallmarks of classic natural product synthesis [19–29], with some exceptions [30]. Investigations into genes potentially involved in toxin synthesis have had some success [31–33], most notably in the separation of genes involved in natural product synthesis from the analogous synthesis of lipids [34–37] and the identification of multi-domain genes [38–40]. These multi-domain genes can then be used to further bin single domain genes into functional groups, although from here the waters become quite muddy with uneven gene copy numbers and the unprecedented duplication of genes related to lipid synthesis [41]. Thus, in many ways, sequence analysis has reached its limits in its ability to shed light on the synthesis of dinoflagellate natural products.

The aim of our project is to extend the current sequence-based knowledge into a biochemical based understanding of natural product synthesis by expressing dinoflagellate proteins in a heterologous system. An attractive target is the carrier protein called the thiolation domain that is activated by the attachment of the phosphopantetheinate group of coenzyme A by a phosphopantetheinyl transferase creating a free thiol moiety. This is the first rate limiting step in natural product synthesis and provides the substrate upon which the actual anabolism is performed by all of the catalytic enzymes. Generally, the activation of a thiolation domain by any phosphopantetheinyl transferase is highly specific and separates specific biosynthetic pathways, although the actual transfer of a phosphopantetheinyl group is not required for recognition of the transferase to a thiolation domain [42]. The thiolation domains of dinoflagellates can be readily separated into two main groups indicative of lipids and natural products [41]. Although the number of thiolation domains can total above one-hundred, the number of phosphopantetheinyl transferase activators is no more than three [43]. In addition, these activators have been expressed in *E. coli* [43] along with the indigoidine synthesis gene BpsA from *Streptomyces lavendulae* [44] that has been placed into an expression vector and characterized previously [45]. The rationale is that, if a given phosphopantetheinyl transferase can activate the thiolation domain of the BpsA reporter, then indigoidine will be produced. This pairing of activator and thiolation domain is a common method for determining specificity [46,47] and has been performed in some protists with a surprising promiscuity not found in bacteria and fungi [48,49]. This study advances previous work by replacing the thiolation domain of the BpsA reporter with several different dinoflagellate sequences to allow for the pairing of each activator with a multitude of potential phosphopantetheination sites. Although the integration of dinoflagellate sequence into the bacterially derived reporter was largely successful, there were several observations that led to the conclusion that this is qualitative only and that several artifacts of heterologous expression need to be overcome in future studies.

2. Materials and Methods

2.1. Reporter Modification

The BpsA reporter described in Owen et al. [45] was kindly obtained from the Ackerley lab at the University of Victoria in Wellington, New Zealand. The region encompassing the thiolation domain was amplified with the primers "BpsA_outF2" and "BpsA_outR2" listed in Table 1 at 500 nm final concentration and 10 µg of vector template using the Phusion high fidelity polymerase (New England Biolabs, Cambridge, MA, USA) as follows: Initial denaturation at 98 °C for 2′; followed by 40 cycles of denaturation at 95 °C for 15 s, annealing at 58 °C for 20″, and extension at 72 °C for 1′ 30″; and polishing at 72 °C for 5′. The amplified product, termed "BpsA_insert0", was purified and sequenced at the BioAnalytical Services Laboratory (BASlab) at the Institute for Marine and Environmental Science in Baltimore MD on an Applied Biosystems 3130 XL. This sequence was used to design the remaining primers in Table 1 to insert a HindIII site in the 3′ end of the thiolation domain and an AflII site in the 5′ end as described in the primer name. The insertions result in the shift of arginine to a lysine at the HindIII site.

Table 1. Synthetic primers and inserts used in reporter modification.

Primers			
Primer Name	**Sequence 5′:3′**	**Length**	**Annealing °C**
BpsA_outF2	TCCAGCACCTGATGATGAAC	20	58.4
BpsA_outR2	CTGGATGCCGTAGAACGAG	19	59.5
BpsAhindiiiR1	GACGCCAAGCTTCGCGTTGAGCTCGCGGACGAGGCCGACGGCGATCAGCGA	51	91.1
BpsAhindiiiF1	CAACGCGAAGCTTGGCGTCTCCCTGCCGCTGCAGAGCGTCCTGGAGTCC	49	89.6
BpsAafliiR1	CTCGCGCTTAAGGGCCTTCTCCCAGACCGCCGCGATCTCCTTCTCCGT	48	88.5
BpsAafliiF1	AGAAGGCCCTTAAGCGCGAGAACGCCTCCGTCCAGGACGACTTCTTCG	48	86.4
Inserts			
Insert Name	Sequence 5′:3′		Binding Site Amino Acid
3KS_1	GAATCGGGCATGGACTCAAAAGCAGCCCTTGTTCTG		**ESG**MDSKAALVL
3KS_2	GAATTGGGCTTAGATTCTTTGTCCGGCGTTGAATTT		ELGLDSLSGVEF
3KS_3	GAAAGCGGAATTGATTCCTTGTCTGCAGTAGAGTTT		ESGIDSLSAVEF
3KS_4	GAGAGTGGCATGGACTCATTATCTGCCGTCGAGTTT		ESGMDSLSAVEF
BurA_1	GCT TCA GGT GCA GAA TCT ATC GCT GTC GTG GGC GTG		ASGAESIAVVGV
BurA_2	CAA TTA GGA TTA GAC AGC TTG GAA ACC GTT CAA CTG		QLGLDSLETVQL
ZmaK_1	GAA ATC GGT GGG CAC TCG CTG TTA GCA ATG AAA CTT		EIGGHSLLAMKL
ZmaK_2	GAT GCC GGG TTA GAT AGC TTA TCC TTA ATT AGC TTA		DAGLDSLSLISL
5′ Linker †	AGAAGGCCCTTAAGCGCGAGAACGCCTCCGTCCAGGACGACTTCTTC		
3′ Linker †	GTCCGCGAGCTCAACGCGAAGCTTGGCGTCTCCCTGCCGCTG		

"†" denotes common linkers to all other inserts and were placed at the 5′ and 3′ ends during synthesis as indicated. The "3KS_1" sequence shown in bold is the wild type sequence included to ensure consistent insert size.

The HindIII and AflII sites were incorporated into the vector in a two-stage process. First, the HindIII site was created via two amplifications using "BpsA_outF2" with "BpsA_hindiiiR1" and "BpsA_outR2" with "BpsA_hindiiiF1" with the same reaction conditions as the thiolation domain amplification. The resultant products were purified using a DNA Clean and Concentrate-5 kit from Zymo research (Irvine, CA, USA) and eluted into 10 uL of distilled deionized water. Approximately 2.5 µg of product was digested with the HindIII-HF restriction enzyme from New England Biolabs for four hours at 37 °C, separated on an ethidium bromide impregnated 1% agarose gel in 0.5× TBE at 15 V/cm for 50 min, excised under ultraviolet illumination, and purified using a Monarch DNA Gel Extraction kit from New England Biolabs (Cambridge, MA, USA) as directed. The two digested fragments were then combined and ligated using a T4 ligase from Promega (Hercules, CA, USA) overnight at 18 °C. This product was then used as template for the second stage amplification using primers "BpsA_outF2" with "BpsA_afliiR1" and "BpsA_outR2" with "BspA_afliiF1" using the same conditions as the HindIII site amplification. This was purified, digested with AflII restriction enzyme from New England Biolabs, agarose gel purified, and combined and ligated in the same manner as the HindIII products resulting in "BpsA_insert1" (Figure 1).

Shown above is the BpsA gene from *Streptomyces lavendulae* originally published in Takahashi et al. [44] showing each of the domains with the thiolation domain marked with a "T". The region surrounded by a blue box is expanded in the bottom showing the thiolation domain and the phosphopantetheinate transferase binding site as red boxes. The existing NotI and FspI restriction as well as the introduced AflII and HindIII sites are shown in red text. The primers used to isolate this region and attach the novel restriction sites are shown as green arrows above with the primer direction indicated by the arrow direction.

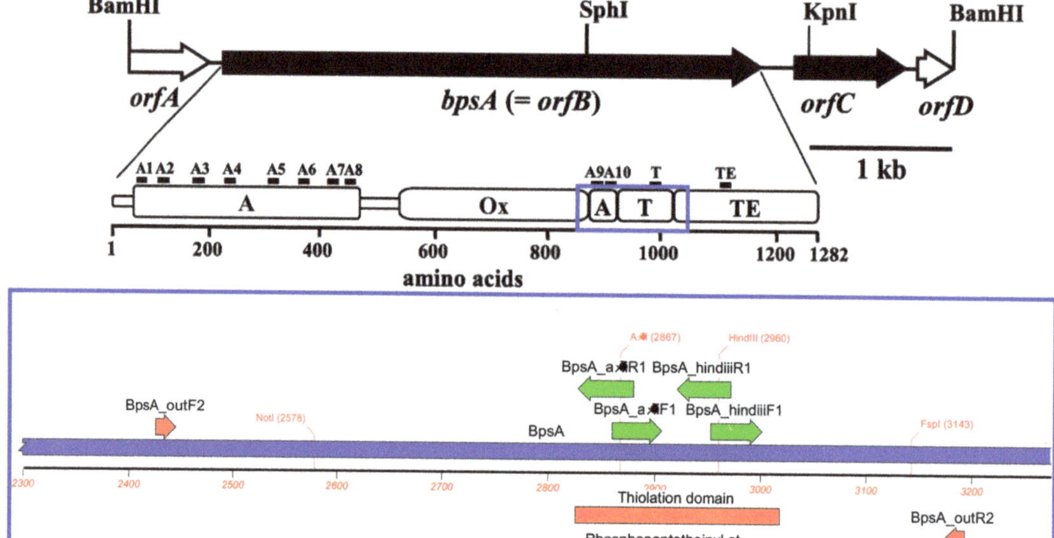

Figure 1. A modification of the BpsA to allow the insertion of dinoflagellate thiolation domain sequences.

BpsA_insert1 was amplified using the same conditions as the original thiolation domain and purified using the DNA Clean and Concentrate-5 kit. This product as well as the original BpsA vector were double digested with the NotI-HF and FspI restriction enzymes from New England Biolabs at 37 °C overnight in cutsmart buffer followed by agarose gel purification and ligation as with the HindIII and AflII amplicons resulting in the BpsA2.1 vector. This was amplified using the Templiphi 100 kit from Cytiva (Marlborough, MA, USA) and cloned into *E. coli* JM109 from Promega (Hercules, CA, USA) according to the manufacturer's directions. A selection of the resultant colonies was grown and the plasmid extracted for co-expression with each of the PPTases from *Amphidinium carterae* as in [43] to confirm activity.

2.2. Thiolation Domain Insertion and Co-Expression

The natural product associated thiolation domains [41] in three multi-domain transcripts (Figure 2) [37,38,41] from *A. carterae* were chosen for complementation in *E. coli* with the three *A. carterae* phosphopantetheinyl transferases (PPTases) that could activate them (Figure 3). These were termed "3KS" for the three ketosynthase domains present, "BurA" for its similarity in sequence and domain arrangement to the BurA gene in *Burkholderia species* [16], and "ZmaK" for the sequence similarity of the dinoflagellate adenylation domain in this transcript to the *Bacillus cereus* adenylation domain in the ZmaK cluster [17]. Each individual thiolation domain was named according to the transcript it was derived from the followed by a numeral indicating the order from 5′ to 3′ in the transcript, e.g., "3KS3" would be the third thiolation domain in the three ketosynthase domain containing transcript. The PPTase binding site amino acid sequence (Table 1) of each thiolation domain was codon optimized for expression in *E. coli* and ordered as an oligonucleotide from Integrated DNA Technologies (Coralville, IA, USA).

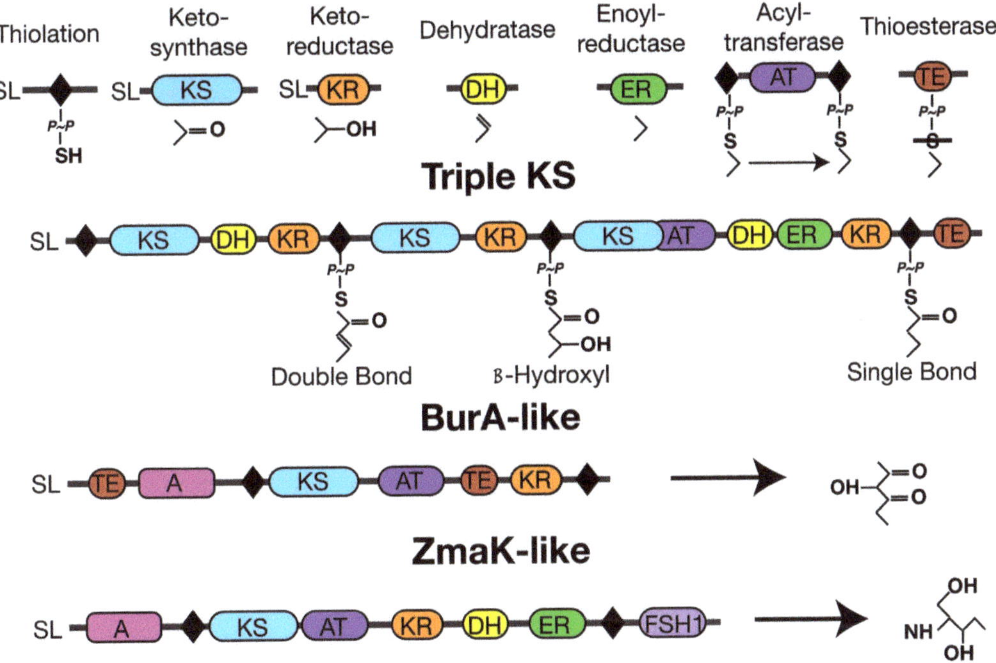

Figure 2. Domain arrangement of *A. carterae* transcripts containing thiolation domains used in this study.

Individual modular synthase domains are shown at the top with example products for their reaction. In addition, Adenylation (A) and FSH1 serine hydrolases (FSH1) are shown for the multi-domain transcripts with examples of potential products included. The phosphopantetheinate group is shown as "P~P" with a single bind to a sulfur. "SL" refers to the dinoflagellate spliced leader sequence and is present if a spliced leader sequence has been verified.

Each oligonucleotide was synthesized with common linker sequences containing the AflII and HindIII restriction sites in the BpsA2.1 plasmid, one for the 5′ end, and one for the 3′ end (Table 1). Thus, each oligonucleotide consisted of the 5′ linker followed by the unique thiolation domain sequence and then the 3′ linker.

For each thiolation domain, the synthetic oligonucleotide as well as the BpsA2.1 plasmid were double digested with HindIII and AflII overnight at 37 °C in cutsmart buffer followed by agarose gel purification using a Monarch DNA Gel Extraction kit from New England Biolabs. The cut insert and plasmid were combined and ligated with a T4 ligase (Promega) at 18 °C overnight. Each ligated plasmid was amplified with the Templiphi 100 kit from Cytiva and cloned into *E. coli* JM109 from Promega according to the manufacturer's directions. JM109 clones were sequenced to verify the presence of the dinoflagellate insert in the plasmid followed by alkaline extraction [51]. Plasmids were then cloned into chemically competent BL21(DE3) *E. coli* (Thermo Fisher, Waltham, MA, USA) along with one of the three PPTase activators (Figure 3) from *A. carterae* in a separate pet-20b plasmid [43] according to the directions for the competent cells and plated onto LB agar containing 100 µg/mL carbenicillin and 50 µg/mL spectinomycin (Sigma Aldrich, St. Louis, MO, USA) at 37 °C. Additionally, each PPTase vector and the thiolation domain vectors were individually cloned into BL21(DE3) to assess protein expression. The vectors for the PPTases were chosen to have a different replication sequence than the reporter to avoid conflicts during growth. Colonies were picked, grown in liquid media containing antibiotics overnight at 37 °C and stored at −80 °C with glycerol added to a final concentration of 12% *v/v*. For assessment of protein expression, glycerol stocks were used to inoculate 10 mL of LB in a 250 mL Erlenmeyer with appropriate antibiotics and grown overnight at 37 °C with shaking at 250 rpm. This was then diluted into 500 mL of LB media with antibiotics in a 2000 mL Erlenmeyer followed by a reduction of temperature to 30 °C and growth for 3 h with shaking. Protein expression was induced by the addition of 500 µL of 0.1 M IPTG followed by incubation at 25 °C for 3 h with shaking. Cells were spun at 10,000× *g* for 10 min at 4 °C, and the media was decanted. Cells were suspended in 20 mL of PBS at 4 °C with a bacterial protease inhibitor (Sigma Aldrich, St. Louis, MO, USA), and proteins were extracted in a French press at 20,000 local PSI followed by centrifugation at 10,000× *g* for 10 min at 4 °C to separate soluble and insoluble material. Insoluble proteins were recovered from the pellet by the addition of 6 M urea in equal volume to the supernatant. Heterologous proteins were purified with a 1 mL HiTrap Talon crude column (Cytiva, Marlborough, MA, USA) on an AKTA chromatography system with elution into 50 mM Tris with 250 mM imidazole. Proteins were separated by SDS-PAGE electrophoresis with 4–12% bis-tris gels (ThermoFisher, Waltham, MA, USA) and imaged with Imperial Coomassie stain (ThermoFisher, Waltham, MA, USA).

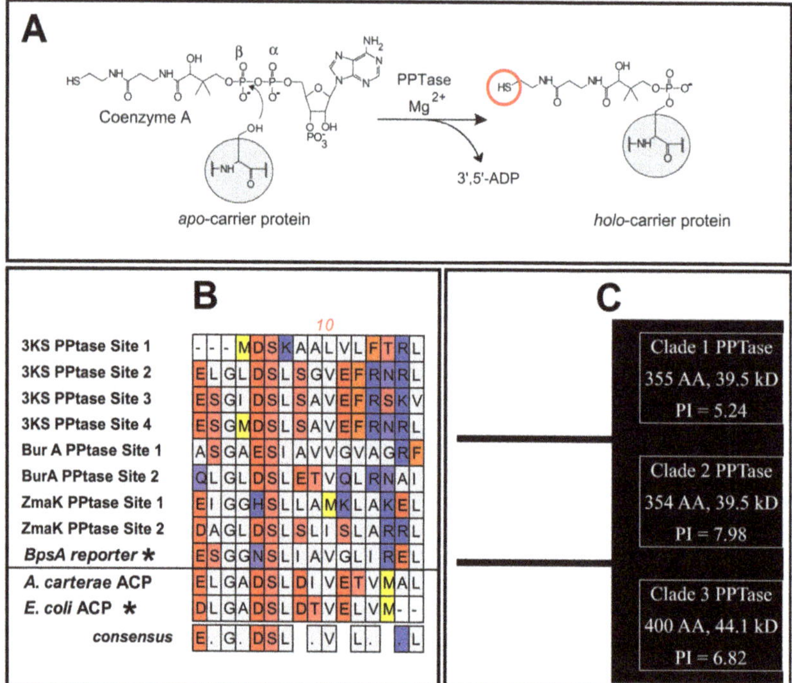

Figure 3. A mechanism of phosphopantetheination and the dinoflagellate thiolation domains used in this study; (**A**) a diagram of the phosphopantetheination reaction from Finking et al. 2002 [50] showing the phosphopantetheinate arm of coenzyme A attachment to the serine of a carrier protein or domain resulting in a free thiol group (red circle); (**B**) the amino acid sequences of the thiolation domains from *A. carterae* used in this study except those marked with a "*" that are from the *S. lavendulae* isolated BpsA gene and the acyl carrier protein (ACP) from *E. coli*. Sequences above the line are theorized to be involved in natural product synthesis while those below the line are for lipid synthesis; (**C**) the predicted folding for the three phosphopantetheinate transferases from *A. carterae*.

The *E. coli* clones containing one of the three *A. carterae* PPTases and one of the eight BpsA reporters with dinoflagellate thiolation domain sequence were each grown onto agar plates containing "autoinduction" media [52]. Colonies were grown at 25 °C for 48 h to allow for growth, protein expression, and indigoidine production. The plates were photographed, and each colony was assessed for dye production by measurement of grayscale density using image J (https://imagej.net/, accessed on 1 December 2018) with the space in between colonies as a baseline for background subtraction.

3. Results

3.1. Construct Generation and Domain Insertion

Following the generation of the BpsA2.1 vector with restriction sites flanking the phosphopantetheinyl transferase (PPTase) binding site of the thiolation domain, each of the eight dinoflagellate thiolation domain oligonucleotides were successfully inserted and verified by Sanger sequencing in both directions (not shown). Although each of the PPTases from *Amphidinium carterae* have previously been shown to interact with the wild type BpsA vector when co-expressed in *E. coli* [43], independent verification of protein production showed very different expression patterns for each of the three PPTases when expressed individually without the BpsA protein (Figure 4). In general, PPTase 3 showed high expression with protein in both the soluble fraction and the insoluble fraction recovered with 6 M urea following lysis of the *E. coli* host by French press. PPTase 2, however, was only visible in the insoluble fraction, and PPTase 1 had low expression in general.

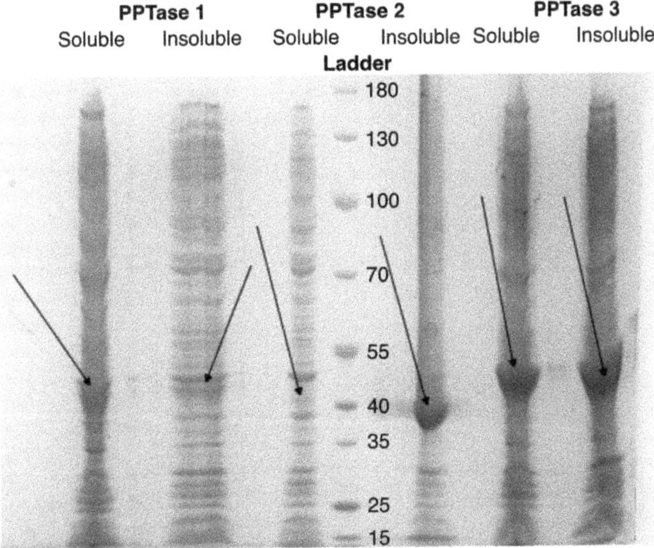

Figure 4. Soluble and insoluble lysates from *E. coli* following induction of phosphopantetheinyl transferase expression.

An SDS-PAGE gel is shown for three *E. coli* clones containing the three *Amphidinium carterae* phosphopantetheinyl transferases following induction of protein expression with IPTG. Both the soluble (supernatant following French press isolation) and insoluble (Proteins retrieved from the pellet with 6 M urea) fractions are shown with arrows indicating the expected size of each protein based on the molecular weight marker designated as "Ladder" with kiloDaltons indicated.

Each of the constructs produced visible protein following his-tag purification (Figure 5) —in contrast to PPTase, the activators where PPTase 2 was not present in the soluble fraction in appreciable amounts. In order to explain how PPTase 2, which was previously shown to activate the wild type BpsA reporter [43], can function despite low apparent soluble production in *E. coli*, co-expression of PPTase 2 with both the BpsA2.1 vector without a heterologous insert as well as with the ZmaK1 insert was his-tag purified (Figure 6). The recoverable amount of the PPTase 2 activator as well as its substrate BpsA protein were higher in the original vector compared to the ZmaK1 insert containing vectors where both the reporter and the PPTase 2 activator were abundant in low amounts.

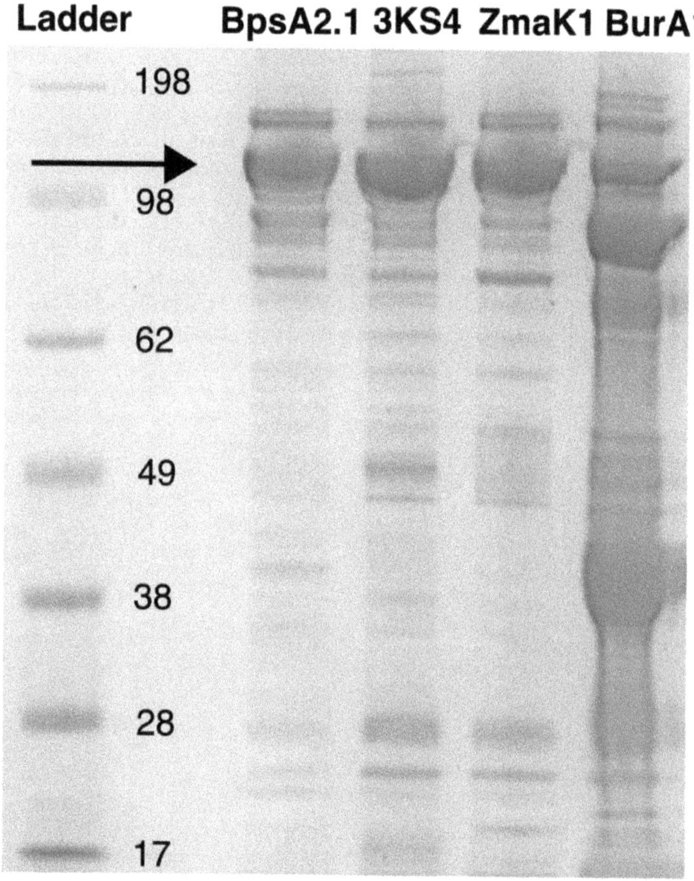

Figure 5. His-tag purified BpsA reporter.

An SDS-PAGE gel is shown for the BpsA2.1 reporter with the standard sequence as well as one each of the triple-KS, ZmaK, and BurA inserts loaded with equivalent total protein. The size marker is shown on the left designated "Ladder" and an arrow shows the expected reporter size. The BurA1 protein was concentrated prior to imaging and shows several breakdown products.

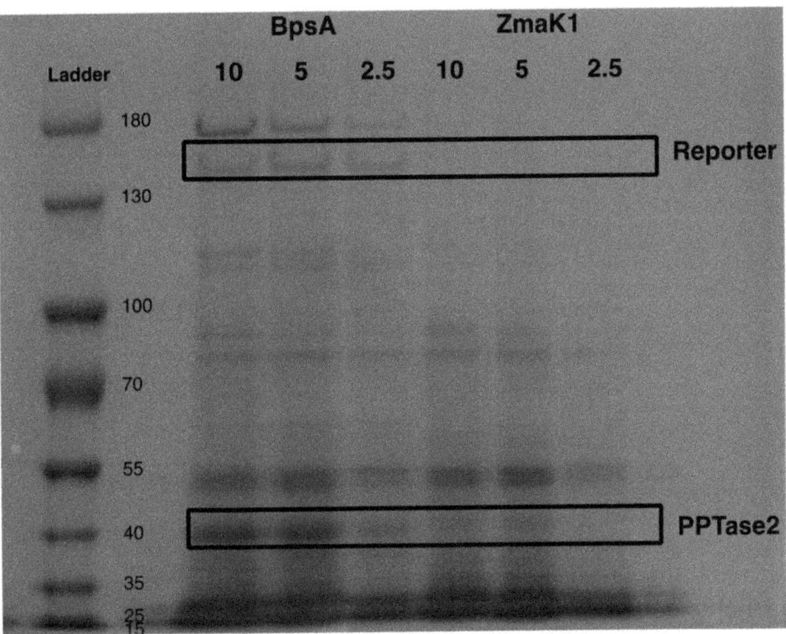

Figure 6. PPtase2 expression with BpsA reporter standard insert and ZmaK insert. An SDS-PAGE gel is shown for a co-expression of PPTase 2 with either the standard BpsA2.1 sequence or with the ZmaK1 insert following French press lysis and removable of insoluble material by centrifugation. The expected sizes for the reporter BpsA protein as well as the PPTase protein are highlighted with black boxes according to the expected size shown on the left with the size standard marked as "Ladder". The load volumes are shown at the top of each well in microliters from equivalent *E. coli* cultures.

3.2. Indigoidine Production

Following growth on autoinduction plates, co-expression of each of the PPTase activators with one of the BpsA2.1 vectors containing either the modified wild-type sequence or a dinoflagellate sequence insert resulted in similar growth for all colonies but indigoidine production in only some colonies (Figure 7). Background subtracted values show higher indigoidine production for the BpsA2.1 vector without inserts as well as with the triple KS inserts but not the ZmaK or BurA inserts with the exception of the combination of BurA2 and PPTase 3. In addition, the PPTase 2 activator pairings yielded consistently lower indigoidine production than PPTase 1 or 3 with the exception of the ZmaK2 insert that had low indigoidine production in all cases but was highest with PPTase 2. Other than the ZmaK2 insert, all BpsA pairings with the PPTase 3 activator resulted in higher indigoidine production relative to the PPTase 1 or 2 activators. Indigoidine production was also performed with each insert along with the PcpS gene, a bacterial PPTase from *Pseudomonas aeruginosa*, and a common control gene for phosphopantetheination, but indigoidine was only produced in appreciable amounts with the reporter without dinoflagellate inserts compared to low or almost no production with the dinoflagellate sequences (Supplementary Figure S1).

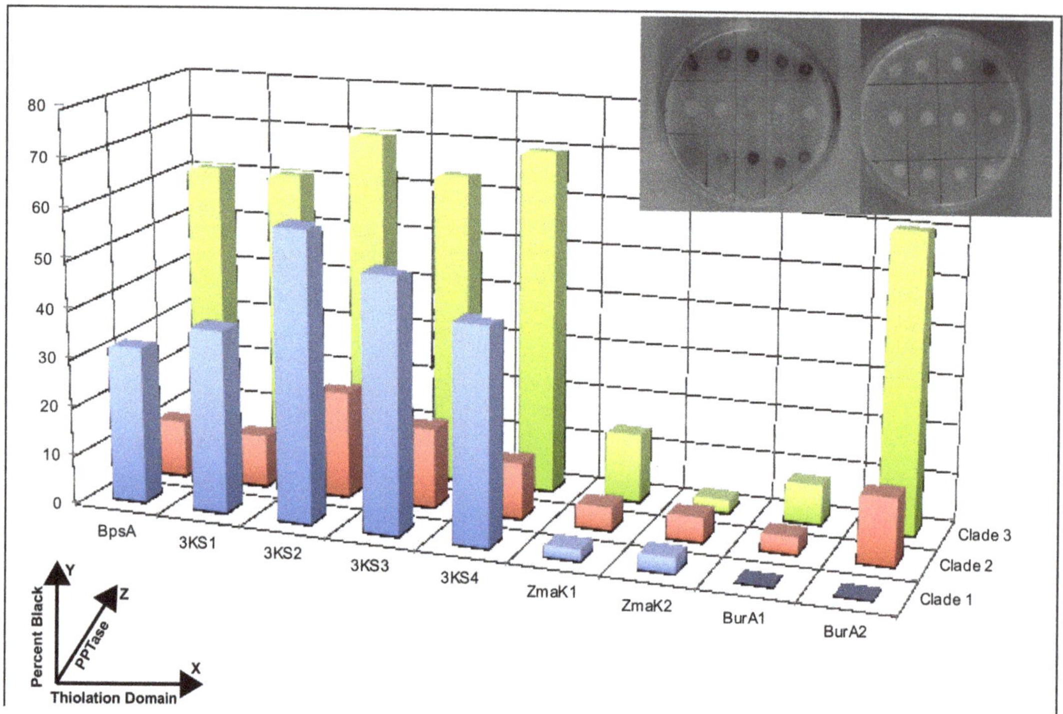

Figure 7. Indigoidine synthesis in *E. coli* from the co-expression of dinoflagellate phosphopantetheinyl transferases and the BpsA gene with a dinoflagellate thiolation domain.

The graph shows the relative darkness of each colony to a black pixel on the *y*-axis resulting from the production of indigoidine in *E. coli* upon the co-expression of the BpsA gene containing a dinoflagellate thiolation domain and a dinoflagellate phosphopantetheinyl transferase (PPTase) on separate vectors. The thiolation domain is indicated on the *x*-axis including wild type sequence (BpsA), as well as the *Amphidinium carterae* triple KS (KS), ZmaK-like (ZmaK), and BurA-like (BurA) transcript sequences with a numeral indicating the particular thiolation domain from the N to C terminus. The *z*-axis indicates which clade of PPTase sequence was co-expressed from Williams et al. [43]. The actual plates with induced expression are shown in the upper right in the same orientation as the graph *x* and *z*-axes for reference.

4. Discussion

Genetic tools such as knockdowns, knockouts, and knockins can be powerful means to determine gene function by looking for phenotype changes following a change in the expression of that gene. While these techniques are well established in many bacteria, fungi, and vertebrates, there are many branches of the tree of life where genetic techniques are not well developed for a variety of reasons. Protists, in particular, representing a huge amount of eukaryotic evolutionary diversity, have lagged behind, although new sequencing technologies have provided a much-needed boost in investigative power [53], resulting in several partial genomes for dinoflagellates in particular [54–58]. Some genetic modifications of dinoflagellates have been successful for the purpose of harnessing the unique lipids of marine protists [59] focusing on knockdowns and knockouts [60,61] but also to investigate the unique biology of dinoflagellates [62]. Success using these techniques is generally limited due to the high copy number of many dinoflagellate genes [18], especially for precision techniques such as CRISPR/Cas9. A gene knock in has been successful for a dinoflagellate chloroplast [63], but gene addition to the nucleus has remained elusive due to the post-transcriptional nature of dinoflagellate gene expression [64–66], making traditional promoter driven approaches unusable. Heterologous expression of dinoflagellate genes in more tractable systems is one way of getting around the biological difficulties of dinoflagellate genetics. This would allow for more direct biochemical analysis of dinoflagellate proteins and has been surprisingly successful with the documented expression of dinoflagellate proteins in mammalian cells [67].

In many ways, this study is a bridge between dinoflagellate biology, focusing on ecology and species diversity, and natural product research, a biochemical approach to discovering and harnessing useful compounds. This marriage seems a foregone conclusion given that dinoflagellates make compounds that affect human health negatively [4] but potentially positively as therapeutic agents, with neosaxitoxin being the first phycotoxin to be used clinically [68]. The difficult biology of dinoflagellates has made this area of research slow-moving although with the production of the polyunsaturated fatty acid DHA from *Crypthecodinium cohnii* [69] and the subsequent efforts for genetic engineering [60] being a notable exception. On the other hand, the natural product world has a great wealth of experience to draw from. Domain replacement has been a common technique to identify substrate specificity or produce novel compounds [70–76] and the BpsA gene itself has also been used extensively [77–80]. In the future, techniques such as NMR and MS based omics approaches can further inform biochemical validation.

Elucidating the function of genes involved in the synthesis of dinoflagellate toxins is especially tricky, since, like most natural products, the synthesis is inherently modular in nature [11], resulting in a large copy number of nearly identical genes. The availability of transcriptomes has helped greatly in cataloging and identifying the genes most likely to be involved in toxin synthesis [31,35,37,39]. While the catalytic genes for toxin synthesis are numerous and similar in sequence, the thiolation domains are less numerous and easily split into those likely to make lipids versus natural products by sequence alone [41]. When considering the phosphopantetheinyl transferases (PPTases) that activate these thiolation domains, which have fewer than three types and a low copy number [43], the number of combinations becomes tractable for biochemical analysis. Thus, this study aims to begin a bottom-up approach, where the specificity of a PPTase for a particular pathway, vis-à-vis that pathway's thiolation domains, can be exploited to isolate toxin synthesis pathways once that specificity has been identified, although there appears to be some unusual promiscuity in protists [49].

Technically, this study represents a step forward as the first example of a catalytically active reporter containing dinoflagellate genes. Natural product synthesis can be quite complicated and dinoflagellate toxins are much larger than most natural products [30]. It is also unclear how many natural products dinoflagellates make since efforts have been so heavily focused towards toxins. In bacteria and fungi, the canon that each biosynthetic pathway for a natural product or lipid has a specific PPTase has been a useful tool in identifying and characterizing each pathway. In dinoflagellates, pathways are very difficult to identify because of the lack of gene synteny and a mathematical problem of copy number, e.g., the number of dehydratase domains is frequently lower than the number of enoyl reductases, even though the dehydratase reaction must occur before the reductase reaction [41]. Thus, biochemical methods like those presented here are necessary to proceed further in identifying natural product biosynthetic pathways in dinoflagellates. While not every combination of an insert containing reporter with a PPTase activator produced indigoidine, each insert was able to produce indigoidine with at least one of the PPTases (Figure 7). Some assumptions necessary for semi-quantitative measurements of indigoidine synthetic potential are that each transformed *E. coli* strain has the same plasmid copy number and expresses each protein equivalently. While the former is likely true, given that equivalent amounts of plasmid were used in the transformation, the latter is certainly false given the differential expression based on PPTase type and reporter pairing (Figures 5 and 6). The *E. coli* colonies were allowed to induce expression and produce indigoidine for 48 h, and, not shown here due to a loss of quantitation, would eventually become dark in color for most insert/activator combinations. This immediately leads to the conclusion that all forms of the reporter are capable of producing indigoidine and that there is some possible phosphopantethienation by each dinoflagellate PPTase for any given thiolation domain, another example of promiscuous PPTase binding in protists [49]. This also verifies the prediction that these enzymes transfer the phosphopantetheinate group and not other moieties, as has been shown in rare cases [50]. In terms of efficiency, PPTase 2 generally produced the lowest amount of indigoidine for all inserts in 48 h, with the exception of ZmaK2, although the variability in the production of soluble protein (Figure 6) makes this result almost certainly spurious. However, if the poor performance of PPTase 2 when combined with these natural product associated thiolation domains is accurate, this leaves open the possibility that PPTase 2 may have more favorable interactions with the acyl carrier protein, the thiolation domain responsible for lipid synthesis. One of the biggest issues with this study is that the dinoflagellate acyl carrier protein is nearly identical to the *E. coli* sequence at the insert site (Figure 3). This construct was not able to be expressed in *E. coli* likely due to toxicity. Thus, comparisons between the thiolation domains presented in this study and the acyl carrier protein would be more appropriate for in-vitro based studies or co-expression studies in another host assuming that the issue of toxicity does not still exist. In contrast to the previous study of dinoflagellate PPTases [43], PPTase 1 generally had similar indigoidine production when compared to PPTase 3 instead of the much lower production previously shown. This is likely due to the much longer autoinduction based protein expression in this study compared to the short term IPTG based expression in the 2020 study.

There were some obvious differences observed between the triple-KS inserts and the ZmaK or BurA inserts in terms of the indigoidine produced (Figure 7). The triple-KS inserts had consistently high indigoidine production, and the triple-KS transcript can also be found in the more basal syndinian dinoflagellate *Hematodinium sp.* [81], a parasite of crustaceans. The BurA-like and ZmaK-like genes on the other hand are not found in any syndinian transcriptomes to date and are very similar in sequence to bacterial genes, making horizontal transfer a likely origin. The results presented here may indicate that, at least for the *Amphidinium carterae* PPTases, the ability to activate the BurA and ZmaK inserts is sub-optimal. This was suggested in the Williams et al. 2021 study on the thiolation domains of dinoflagellate domains based on the observation that many of the ZmaK sequences lie outside the cluster of natural product associated domains with the more basal sequence the furthest away, indicating that convergent evolution may be an active force. Thus, PPTases from more distal species of dinoflagellates may be better at activating the BurA and ZmaK insert containing reporters than the *A. carterae* based PPTases used here. It could also be that the BurA and ZmaK based sequences themselves are interfering with the reporter's ability to produce indigoidine given that the thiolation domain acts as an intermediary for all other domains. The BurA and ZmaK sequences may be sterically interfering with the other domains of the BpsA protein, reducing its overall efficiency.

5. Conclusions

The modification and deployment of this indigoidine synthesizing reporter for dinoflagellate genes allows for new approaches to studying genes that may be involved in toxin synthesis. Specifically, methods that can validate suspected interactions during natural product synthesis are crucial in overcoming the limitations of sequence-based annotations. Here, domain substitution has been used to demonstrate the broad substrate recognition of dinoflagellate PPTases that can help explain gene losses and gains throughout dinoflagellate evolution that do not correlate with a loss or gain in toxin synthesis. There are also several not unexpected lessons to be learned from these results. First and foremost is that, despite the huge evolutionary distance between dinoflagellates and the *E. coli* heterologous host, some dinoflagellate genes simply cannot be expressed in this system, likely due to host interactions and resultant toxicity. This is both intriguing and frustrating that genes involved in lipid synthesis can be so conserved in such a dynamic group of organisms like dinoflagellates. Perhaps some comparisons to syndinian dinoflagellates that parasitize and absorb lipids from dinoflagellate hosts can shed light on how these genes function and have been modified over time. A second consideration is the dynamic nature of expression of the PPTases depending on the thiolation domain it is paired with. This renders these results qualitative at best and implies that in vitro methods, while more challenging, are likely more accurate comparisons of PPTase activity. Finally, the lack of indigoidine synthesis for the BpsA vectors containing the ZmaK and BurA inserts is suspicious and may be the result of steric hindrance rather than a lack of PPTase activation. A general conclusion from the activation of all the thiolation domain sequences used here by all three PPTases is that the substrate specificity observed in bacteria and fungi may be the exception in protists rather than the rule. The mechanism that dinoflagellates use to regulate the initiation of lipid and natural product synthesis is an important topic for future study. Targeted knockdowns are another avenue going forward to validate these results since broad substrate recognition shown here does not mean that specific natural product pathways are not physically separated in situ. In addition, knockdowns of acyl transferases and thioesterases that join and terminate major portions of natural product synthesis as proposed in Van Wagoner et al. 2014 [30] can help to isolate specific pathways.

Supplementary Materials: The following supporting information can be downloaded at: https://www.mdpi.com/article/10.3390/microorganisms10040687/s1, Figure S1: Indigoidine Production by PcpS.

Author Contributions: Conceptualization, review and editing by E.W., T.B. and A.P.; Methodology by E.W. and A.P.; Supervision by A.P.; and Formal analysis and writing by E.W. All authors have read and agreed to the published version of the manuscript.

Funding: This research received no external funding.

Institutional Review Board Statement: Not Applicable.

Informed Consent Statement: Not Applicable.

Data Availability Statement: Not applicable.

Acknowledgments: We would like to acknowledge the lab of David Ackerley for the donation of the BpsA containing plasmid and the Sabeena Nazar at the BASlab for sequencing support.

Conflicts of Interest: The authors declare no conflict of interest.

References

1. Deeds, J.R.; Terlizzi, D.E.; Adolf, J.E.; Stoecker, D.K.; Place, A.R. Toxic activity from cultures of *Karlodinium micrum* (=*Gyrodinium galatheanum*) (Dinophyceae)—A dinoflagellate associated with fish mortalities in an estuarine aquaculture facility. *Harmful Algae* **2002**, *1*, 169–189. [CrossRef]
2. Twiner, M.J.; Flewelling, L.J.; Fire, S.E.; Bowen-Stevens, S.R.; Gaydos, J.K.; Johnson, C.K.; Landsberg, J.H.; Leighfield, T.A.; Mase-Guthrie, B.; Schwacke, L.; et al. Comparative analysis of three brevetoxin-associated bottlenose dolphin (*Tursiops truncatus*) mortality events in the Florida Panhandle region (USA). *PLoS ONE* **2012**, *7*, e42974. [CrossRef] [PubMed]
3. Walsh, C.J.; Butawan, M.; Yordy, J.; Ball, R.; Flewelling, L.; de Wit, M.; Bonde, R.K. Sublethal red tide toxin exposure in free-ranging manatees (*Trichechus manatus*) affects the immune system through reduced lymphocyte proliferation responses, inflammation, and oxidative stress. *Aquat. Toxicol.* **2015**, *161*, 73–84. [CrossRef] [PubMed]
4. Wang, D.-Z. Neurotoxins from marine dinoflagellates: A brief review. *Mar. Drugs* **2008**, *6*, 349–371. [CrossRef] [PubMed]
5. Adolf, J.E.; Krupatkina, D.; Bachvaroff, T.; Place, A.R. Karlotoxin mediates grazing by *Oxyrrhis marina* on strains of *Karlodinium veneficum*. *Harmful Algae* **2007**, *6*, 400–412. [CrossRef]
6. Sheng, J.; Malkiel, E.; Katz, J.; Adolf, J.E.; Place, A.R. A dinoflagellate exploits toxins to immobilize prey prior to ingestion. *Proc. Natl. Acad. Sci. USA* **2010**, *107*, 2082–2087. [CrossRef]
7. Colon, R.; Wheater, M.; Joyce, E.J.; Ste Marie, E.J.; Hondal, R.J.; Rein, K.S. The Marine Neurotoxin Brevetoxin (PbTx-2) Inhibits Karenia brevis and Mammalian Thioredoxin Reductases by Targeting Different Residues. *J. Nat. Prod.* **2021**, *84*, 2961–2970. [CrossRef]
8. Chen, W.; Colon, R.; Louda, J.W.; Del Rey, F.R.; Durham, M.; Rein, K.S. Brevetoxin (PbTx-2) influences the redox status and NPQ of *Karenia brevis* by way of thioredoxin reductase. *Harmful Algae* **2018**, *71*, 29–39. [CrossRef]

9. Beld, J.; Sonnenschein, E.C.; Vickery, C.R.; Noel, J.P.; Burkart, M.D. The phosphopantetheinyl transferases: Catalysis of a post-translational modification crucial for life. *Nat. Prod. Rep.* **2014**, *31*, 61–108. [CrossRef]

10. Bentley, R.; Bennett, J.W. Constructing polyketides: From collie to combinatorial biosynthesis. *Annu. Rev. Microbiol.* **1999**, *53*, 411–446. [CrossRef]

11. Khosla, C. Structures and mechanisms of polyketide synthases. *J. Org. Chem.* **2009**, *74*, 6416–6420. [CrossRef]

12. Sieber, S.A.; Marahiel, M.A. Molecular mechanisms underlying nonribosomal peptide synthesis: Approaches to new antibiotics. *Chem. Rev.* **2005**, *105*, 715–738. [CrossRef]

13. Khosla, C.; Kapur, S.; Cane, D.E. Revisiting the modularity of modular polyketide synthases. *Curr. Opin. Chem. Biol.* **2009**, *13*, 135–143. [CrossRef]

14. Rausch, C.; Hoof, I.; Weber, T.; Wohlleben, W.; Huson, D.H. Phylogenetic analysis of condensation domains in NRPS sheds light on their functional evolution. *BMC Evol. Biol.* **2007**, *7*, 78. [CrossRef]

15. Wang, H.; Fewer, D.P.; Holm, L.; Rouhiainen, L.; Sivonen, K. Atlas of nonribosomal peptide and polyketide biosynthetic pathways reveals common occurrence of nonmodular enzymes. *Proc. Natl. Acad. Sci. USA* **2014**, *111*, 9259–9264. [CrossRef]

16. Franke, J.; Ishida, K.; Hertweck, C. Genomics-driven discovery of burkholderic acid, a noncanonical, cryptic polyketide from human pathogenic Burkholderia species. *Angew Chem. Int. Ed. Engl.* **2012**, *51*, 11611–11615. [CrossRef]

17. Kevany, B.M.; Rasko, D.A.; Thomas, M.G. Characterization of the complete zwittermicin A biosynthesis gene cluster from *Bacillus cereus*. *Appl. Environ. Microbiol.* **2009**, *75*, 1144–1155. [CrossRef]

18. Bachvaroff, T.R.; Place, A.R. From stop to start: Tandem gene arrangement, copy number and trans-splicing sites in the dinoflagellate *Amphidinium carterae*. *PLoS ONE* **2008**, *3*, e2929. [CrossRef]

19. Fukatsu, T.; Onodera, K.; Ohta, Y.; Oba, Y.; Nakamura, H.; Shintani, T.; Yoshioka, Y.; Okamoto, T.; ten Lohuis, M.; Miller, D.J.; et al. Zooxanthellamide D, a polyhydroxy polyene amide from a marine dinoflagellate, and chemotaxonomic perspective of the *Symbiodinium* polyols. *J. Nat. Prod.* **2007**, *70*, 407–411. [CrossRef]

20. Ishida, H.; Nozawa, A.; Totoribe, K.; Muramatsu, N.; Nukaya, H.; Tsuji, K.; Yamaguchi, K.; Yasumoto, T.; Kaspar, H.; Berkett, N. Brevetoxin B1, a new polyether marine toxin from the New Zealand shellfish, *Austrovenus stutchburyi*. *Tetrahedron Lett.* **1995**, *36*, 725–728. [CrossRef]

21. Macpherson, G.R.; Burton, I.W.; LeBlanc, P.; Walter, J.A.; Wright, J.L. Studies of the biosynthesis of DTX-5a and DTX-5b by the dinoflagellate *Prorocentrum maculosum*: Regiospecificity of the putative Baeyer-Villigerase and insertion of a single amino acid in a polyketide chain. *J. Org. Chem.* **2003**, *68*, 1659–1664. [CrossRef] [PubMed]

22. Meng, Y.; Van Wagoner, R.M.; Misner, I.; Tomas, C.; Wright, J.L. Structure and biosynthesis of amphidinol 17, a hemolytic compound from *Amphidinium carterae*. *J. Nat. Prod.* **2010**, *73*, 409–415. [CrossRef] [PubMed]

23. Peng, J.; Place, A.R.; Yoshida, W.; Anklin, C.; Hamann, M.T. Structure and absolute configuration of karlotoxin-2, an ichthyotoxin from the marine dinoflagellate *Karlodinium veneficum*. *J. Am. Chem. Soc.* **2010**, *132*, 3277–3279. [CrossRef]

24. Sasaki, M.; Matsumori, N.; Maruyama, T.; Nonomura, T.; Murata, M.; Tachibana, K.; Yasumoto, T. The Complete Structure of Maitotoxin, Part I: Configuration of the C1–C14 Side Chain. *Angew. Chem. Int. Ed. Engl.* **1996**, *35*, 1672–1675. [CrossRef]

25. Satake, M.; Morohashi, A.; Oguri, H.; Oishi, T.; Hirama, M.; Harada, N.; Yasumoto, T. The absolute configuration of ciguatoxin. *J. Am. Chem. Soc.* **1997**, *119*, 11325–11326. [CrossRef]

26. Seki, T.; Satake, M.; Mackenzie, L.; Kaspar, H.F.; Yasumoto, T. Gymnodimine, a new marine toxin of unprecedented structure isolated from New Zealand oysters and the dinoflagellate, *Gymnodinium* sp. *Tetrahedron Lett.* **1995**, *36*, 7093–7096. [CrossRef]

27. Van Wagoner, R.M.; Deeds, J.R.; Satake, M.; Ribeiro, A.A.; Place, A.R.; Wright, J.L. Isolation and characterization of karlotoxin 1, a new amphipathic toxin from *Karlodinium veneficum*. *Tetrahedron Lett.* **2008**, *49*, 6457–6461. [CrossRef]

28. Van Wagoner, R.M.; Deeds, J.R.; Tatters, A.O.; Place, A.R.; Tomas, C.R.; Wright, J.L. Structure and relative potency of several karlotoxins from *Karlodinium veneficum*. *J. Nat. Prod.* **2010**, *73*, 1360–1365. [CrossRef]

29. Wright, J.L.C.; Hu, T.; McLachlan, J.L.; Needham, J.; Walter, J.A. Biosynthesis of DTX-4: Confirmation of a polyketide pathway, proof of a Baeyer—Villiger oxidation step, and evidence for an unusual carbon deletion process. *J. Am. Chem. Soc.* **1996**, *118*, 8757–8758. [CrossRef]

30. Van Wagoner, R.M.; Satake, M.; Wright, J.L. Polyketide biosynthesis in dinoflagellates: What makes it different. *Nat. Prod. Rep.* **2014**, *31*, 1101–1137. [CrossRef]

31. Verma, A.; Barua, A.; Ruvindy, R.; Savela, H.; Ajani, P.A.; Murray, S.A. The Genetic Basis of Toxin Biosynthesis in Dinoflagellates. *Microorganisms* **2019**, *7*, 222. [CrossRef]

32. Snyder, R.V.; Gibbs, P.D.L.; Palacios, A.; Abiy, L.; Dickey, R.; Lopez, J.V.; Rein, K.S. Polyketide synthase genes from marine dinoflagellates. *Mar. Biotechnol.* **2003**, *5*, 1–12. [CrossRef]

33. Beedessee, G.; Hisata, K.; Roy, M.C.; Van Dolah, F.M.; Satoh, N.; Shoguchi, E. Diversified secondary metabolite biosynthesis gene repertoire revealed in symbiotic dinoflagellates. *Sci. Rep.* **2019**, *9*, 1204. [CrossRef]

34. Kohli, G.S.; John, U.; Figueroa, R.I.; Rhodes, L.L.; Harwood, D.T.; Groth, M.; Bolch, C.J.; Murray, S.A. Polyketide synthesis genes associated with toxin production in two species of *Gambierdiscus* (Dinophyceae). *BMC Genom.* **2015**, *16*, 410. [CrossRef]

35. Kohli, G.S.; John, U.; Van Dolah, F.M.; Murray, S.A. Evolutionary distinctiveness of fatty acid and polyketide synthesis in eukaryotes. *ISME J.* **2016**, *10*, 1877–1890. [CrossRef]

36. Meyer, J.M.; Rödelsperger, C.; Eichholz, K.; Tillmann, U.; Cembella, A.; McGaughran, A.; John, U. Transcriptomic characterisation and genomic glimps into the toxigenic dinoflagellate *Azadinium spinosum*, with emphasis on polykeitde synthase genes. *BMC Genom.* **2015**, *16*, 27. [CrossRef]

37. Van Dolah, F.M.; Kohli, G.S.; Morey, J.S.; Murray, S.A. Both modular and single-domain Type I polyketide synthases are expressed in the brevetoxin-producing dinoflagellate, *Karenia brevis* (Dinophyceae). *J. Phycol.* **2017**, *53*, 1325–1339. [CrossRef]

38. Bachvaroff, T.R.; Williams, E.P.; Jagus, R.; Place, A.R. A cryptic noncanonical multi-module PKS/NRPS found in dinoflagellates. In Proceedings of the 16th International Conference on Harmful Algae, Wellington, New Zealand, 27–31 October 2014; pp. 101–104.

39. Kohli, G.S.; Campbell, K.; John, U.; Smith, K.F.; Fraga, S.; Rhodes, L.L.; Murray, S.A. Role of Modular Polyketide Synthases in the Production of Polyether Ladder Compounds in Ciguatoxin-Producing *Gambierdiscus polynesiensis* and *G. excentricus* (Dinophyceae). *J. Eukaryot Microbiol.* **2017**, *64*, 691–706. [CrossRef]

40. Van Dolah, F.M.; Morey, J.S.; Milne, S.; Ung, A.; Anderson, P.E.; Chinain, M. Transcriptomic analysis of polyketide synthases in a highly ciguatoxic dinoflagellate, *Gambierdiscus polynesiensis* and low toxicity *Gambierdiscus pacificus*, from French Polynesia. *PLoS ONE* **2020**, *15*, e0231400. [CrossRef]

41. Williams, E.P.; Bachvaroff, T.R.; Place, A.R. A Global Approach to Estimating the Abundance and Duplication of Polyketide Synthase Domains in Dinoflagellates. *Evol. Bioinform. Online* **2021**, *17*, 11769343211031871. [CrossRef]

42. Bunkoczi, G.; Pasta, S.; Joshi, A.; Wu, X.; Kavanagh, K.L.; Smith, S.; Oppermann, U. Mechanism and substrate recognition of human holo ACP synthase. *Chem. Biol.* **2007**, *14*, 1243–1253. [CrossRef] [PubMed]

43. Williams, E.P.; Bachvaroff, T.R.; Place, A.R. The Phosphopantetheinyl Transferases in Dinoflagellates. In Proceedings of the 18th International Conference on Harmful Algae, Harmful Algae 2018-from Ecosystems to Socioecosystems, Nantes, France, 21–26 October 2018; pp. 176–180.

44. Takahashi, H.; Kumagai, T.; Kitani, K.; Mori, M.; Matoba, Y.; Sugiyama, M. Cloning and characterization of a *Streptomyces* single module type non-ribosomal peptide synthetase catalyzing a blue pigment synthesis. *J. Biol. Chem.* **2007**, *282*, 9073–9081. [CrossRef] [PubMed]

45. Owen, J.G.; Copp, J.N.; Ackerley, D.F. Rapid and flexible biochemical assays for evaluating 4'-phosphopantetheinyl transferase activity. *Biochem. J.* **2011**, *436*, 709–717. [CrossRef] [PubMed]

46. Geerlof, A.; Lewendon, A.; Shaw, W.V. Purification and Characterization of Phosphopantetheine Adenylyltransferase from *Escherichia coli*. *J. Biol. Chem.* **1999**, *274*, 27105–27111. [CrossRef]

47. Murugan, E.; Kong, R.; Sun, H.; Rao, F.; Liang, Z.X. Expression, purification and characterization of the acyl carrier protein phosphodiesterase from *Pseudomonas Aeruginosa*. *Protein. Expr. Purif.* **2010**, *71*, 132–138. [CrossRef]

48. Cai, X.; Herschap, D.; Zhu, G. Functional characterization of an evolutionarily distinct phosphopantetheinyl transferase in the apicomplexan *Cryptosporidium parvum*. *Eukaryot Cell* **2005**, *4*, 1211–1220. [CrossRef]

49. Sonnenschein, E.C.; Pu, Y.; Beld, J.; Burkart, M.D. Phosphopantetheinylation in the green microalgae Chlamydomonas reinhardtii. *J. Appl. Phycol.* **2016**, *28*, 3259–3267. [CrossRef]

50. Finking, R.; Solsbacher, J.; Konz, D.; Schobert, M.; Schafer, A.; Jahn, D.; Marahiel, M.A. Characterization of a new type of phosphopantetheinyl transferase for fatty acid and siderophore synthesis in *Pseudomonas aeruginosa*. *J. Biol. Chem.* **2002**, *277*, 50293–50302. [CrossRef]

51. Sambrook, J.; Fritsch, E.F.; Maniatis, T. *Molecular Cloning: A Laboratory Manual*; Cold Spring Harbor Laboratory Press: Cold Spring Harbor, NY, USA, 1989; p. 654.

52. Studier, F.W. Protein production by auto-induction in high density shaking cultures. *Protein. Expr. Purif.* **2005**, *41*, 207–234. [CrossRef]

53. Faktorová, D.; Nisbet, R.E.R.; Fernández Robledo, J.A.; Casacuberta, E.; Sudek, L.; Allen, A.E.; Ares, M.; Aresté, C.; Balestreri, C.; Barbrook, A.C.; et al. Genetic tool development in marine protists: Emerging model organisms for experimental cell biology. *Nat. Methods.* **2020**, *17*, 481–494. [CrossRef]

54. Beedessee, G.; Kubota, T.; Arimoto, A.; Nishitsuji, K.; Waller, R.F.; Hisata, K.; Yamasaki, S.; Satoh, N.; Kobayashi, J.; Shoguchi, E. Integrated omics unveil the secondary metabolic landscape of a basal dinoflagellate. *BMC Biol.* **2020**, *18*, 139. [CrossRef]

55. Stephens, T.G.; González-Pech, R.A.; Cheng, Y.; Mohamed, A.R.; Burt, D.W.; Bhattacharya, D.; Ragan, M.A.; Chan, C.X. Genomes of the dinoflagellate *Polarella glacialis* encode tandemly repeated single-exon genes with adaptive functions. *BMC Biol.* **2020**, *18*, 56. [CrossRef]

56. Aranda, M.; Li, Y.; Liew, Y.J.; Baumgarten, S.; Simakov, O.; Wilson, M.C.; Piel, J.; Ashoor, H.; Bougouffa, S.; Bajic, V.B.; et al. Genomes of coral dinoflagellate symbionts highlight evolutionary adaptations conducive to a symbiotic lifestyle. *Sci. Rep.* **2016**, *6*, 39734. [CrossRef]

57. Lin, S.; Cheng, S.; Song, B.; Zhong, X.; Lin, X.; Li, W.; Li, L.; Zhang, Y.; Zhang, H.; Ji, Z.; et al. The *Symbiodinium kawagutii* genome illuminates dinoflagellate gene expression and coral symbiosis. *Science* **2015**, *350*, 691–694. [CrossRef]

58. Shoguchi, E.; Shinzato, C.; Kawashima, T.; Gyoja, F.; Mungpakdee, S.; Koyanagi, R.; Takeuchi, T.; Hisata, K.; Tanaka, M.; Fujiwara, M.; et al. Draft assembly of the *Symbiodinium minutum* nuclear genome reveals dinoflagellate gene structure. *Curr. Biol.* **2013**, *23*, 1399–1408. [CrossRef]

59. Leblond, J.D.; Evans, T.J.; Chapman, P.J. The biochemistry of dinoflagellate lipids, with particular reference to the fatty acid and sterol composition of a *Karenia brevis* bloom. *Phycologia* **2003**, *42*, 324–331. [CrossRef]

60. Diao, J.; Song, X.; Zhang, X.; Chen, L.; Zhang, W. Genetic Engineering of Crypthecodinium cohnii to Increase Growth and Lipid Accumulation. *Front Microbiol.* **2018**, *9*, 492. [CrossRef]
61. Tuttle, R.C.; Loeblich, A.R. Genetic recombination in the dinoflagellate *Crypthecodinium cohnii*. *Science* **1974**, *185*, 1061–1062. [CrossRef]
62. Yan, T.H.K.; Wu, Z.; Kwok, A.C.M.; Wong, J.T.Y. Knockdown of Dinoflagellate Condensin CcSMC4 Subunit Leads to S-Phase Impediment and Decompaction of Liquid Crystalline Chromosomes. *Microorganisms* **2020**, *8*, 565. [CrossRef]
63. Nimmo, I.C.; Barbrook, A.C.; Lassadi, I.; Chen, J.E.; Geisler, K.; Smith, A.G.; Aranda, M.; Purton, S.; Waller, R.F.; Nisbet, R.E.R.; et al. Genetic transformation of the dinoflagellate chloroplast. *eLife* **2019**, *8*, e45292. [CrossRef]
64. Lidie, K.B.; Ryan, J.C.; Barbier, M.; Van Dolah, F.M. Gene expression in Florida red tide dinoflagellate *Karenia brevis*: Analysis of an expressed sequence tag library and development of DNA microarray. *Mar. Biotechnol.* **2005**, *7*, 481–493. [CrossRef] [PubMed]
65. Morse, D.; Milos, P.M.; Roux, E.; Hastings, J.W. Circadian regulation of bioluminescence in *Gonyaulax* involves translational control. *Proc. Natl. Acad. Sci. USA* **1989**, *86*, 172–176. [CrossRef]
66. Roy, S.; Jagus, R.; Morse, D. Translation and Translational Control in Dinoflagellates. *Microorganisms* **2018**, *6*, 30. [CrossRef] [PubMed]
67. Ma, M.; Shi, X.; Lin, S. Heterologous expression and cell membrane localization of dinoflagellate opsins (rhodopsin proteins) in mammalian cells. *Mar. Life Sci. Technol.* **2020**, *2*, 302–308. [CrossRef]
68. Manríquez, V.; Castro Caperan, D.; Guzmán, R.; Naser, M.; Iglesia, V.; Lagos, N. First evidence of neosaxitoxin as a long-acting pain blocker in bladder pain syndrome. *Int. Urogynecol. J.* **2015**, *26*, 853–858. [CrossRef] [PubMed]
69. Mendes, A.; Reis, A.; Vasconcelos, R.; Guerra, P.; Lopes da Silva, T. *Crypthecodinium cohnii* with emphasis on DHA production: A review. *J. Appl. Phycol.* **2009**, *21*, 199–214. [CrossRef]
70. Jiao, Y.L.; Wang, L.H.; Jiao, B.H.; Wang, S.J.; Fang, Y.W.; Liu, S. Function analysis of a new type I PKS-SAT domain by SAT-EAT domain replacement. *Prikl. Biokhim. Mikrobiol.* **2010**, *46*, 161–165.
71. Wang, H.; Liang, J.; Yue, Q.; Li, L.; Shi, Y.; Chen, G.; Li, Y.Z.; Bian, X.; Zhang, Y.; Zhao, G.; et al. Engineering the acyltransferase domain of epothilone polyketide synthase to alter the substrate specificity. *Microb. Cell Fact.* **2021**, *20*, 86. [CrossRef]
72. Brautaset, T.; Borgos, S.E.; Sletta, H.; Ellingsen, T.E.; Zotchev, S.B. Site-specific mutagenesis and domain substitutions in the loading module of the nystatin polyketide synthase, and their effects on nystatin biosynthesis in *Streptomyces noursei*. *J. Biol. Chem.* **2003**, *278*, 14913–14919. [CrossRef]
73. Calcott, M.J.; Owen, J.G.; Ackerley, D.F. Efficient rational modification of non-ribosomal peptides by adenylation domain substitution. *Nat. Commun.* **2020**, *11*, 4554. [CrossRef]
74. Calcott, M.J.; Ackerley, D.F. Portability of the thiolation domain in recombinant pyoverdine non-ribosomal peptide synthetases. *BMC Microbiol.* **2015**, *15*, 162. [CrossRef]
75. Butz, D.; Schmiederer, T.; Hadatsch, B.; Wohlleben, W.; Weber, T.; Süssmuth, R.D. Module extension of a non-ribosomal peptide synthetase of the glycopeptide antibiotic balhimycin produced by *Amycolatopsis balhimycina*. *Chembiochem* **2008**, *9*, 1195–1200. [CrossRef]
76. Khosla, C.; Ebert-Khosla, S.; Hopwood, D.A. Targeted gene replacements in a *Streptomyces* polyketide synthase gene cluster: Role for the acyl carrier protein. *Mol. Microbiol.* **1992**, *6*, 3237–3249. [CrossRef]
77. Brown, A.S.; Calcott, M.J.; Collins, V.M.; Owen, J.G.; Ackerley, D.F. The indigoidine synthetase BpsA provides a colorimetric ATP assay that can be adapted to quantify the substrate preferences of other NRPS enzymes. *Biotechnol. Lett.* **2020**, *42*, 2665–2671. [CrossRef]
78. Brown, A.S.; Robins, K.J.; Ackerley, D.F. A sensitive single-enzyme assay system using the non-ribosomal peptide synthetase BpsA for measurement of L-glutamine in biological samples. *Sci. Rep.* **2017**, *7*, 41745. [CrossRef]
79. Wehrs, M.; Prahl, J.P.; Moon, J.; Li, Y.; Tanjore, D.; Keasling, J.D.; Pray, T.; Mukhopadhyay, A. Production efficiency of the bacterial non-ribosomal peptide indigoidine relies on the respiratory metabolic state in S. cerevisiae. *Microb. Cell Fact.* **2018**, *17*, 193. [CrossRef]
80. Owen, J.G.; Calcott, M.J.; Robins, K.J.; Ackerley, D.F. Generating Functional Recombinant NRPS Enzymes in the Laboratory Setting via Peptidyl Carrier Protein Engineering. *Cell Chem. Biol.* **2016**, *23*, 1395–1406. [CrossRef]
81. Gornik, S.G.; Cassin, A.M.; MacRae, J.I.; Ramaprasad, A.; Rchiad, Z.; McConville, M.J.; Bacic, A.; McFadden, G.I.; Pain, A.; Waller, R.F. Endosymbiosis undone by stepwise elimination of the plastid in a parasitic dinoflagellate. *Proc. Natl. Acad. Sci. USA* **2015**, *112*, 5767–5772. [CrossRef]

 *microorganisms*

Communication

Transcriptome Analysis of *Durusdinium* Associated with the Transition from Free-Living to Symbiotic

Ikuko Yuyama [1],*, Naoto Ugawa [2] and Tetsuo Hashimoto [2]

[1] Graduate School of Science and Technology for Innovation, Yamaguchi University, 1677-1 Yoshida, Yamaguchi 753-8512, Japan

[2] Faculty of Life and Environmental Sciences, University of Tsukuba, 111 Tennodai, Tsukuba, Ibaraki 305-8577, Japan; ugawa.naoto.sw@alumni.tsukuba.ac.jp (N.U.); hashimoto.tetsuo.gm@u.tsukuba.ac.jp (T.H.)

* Correspondence: yuyamai@gmail.com; Tel.: +81-83-933-5835

Abstract: To detect the change during coral–dinoflagellate endosymbiosis establishment, we compared transcriptome data derived from free-living and symbiotic *Durusdinium*, a coral symbiont genus. We detected differentially expressed genes (DEGs) using two statistical methods (edgeR using raw read data and the Student's *t*-test using bootstrap resampling read data) and detected 1214 DEGs between the symbiotic and free-living states, which we subjected to gene ontology (GO) analysis. Based on the representative GO terms and 50 DEGs with low false discovery rates, changes in *Durusdinium* during endosymbiosis were predicted. The expression of genes related to heat-shock proteins and microtubule-related proteins tended to decrease, and those of photosynthesis genes tended to increase. In addition, a phylogenetic analysis of dapdiamide A (antibiotics) synthase, which was upregulated among the 50 DEGs, confirmed that two genera in the Symbiodiniaceae family, *Durusdinium* and *Symbiodinium*, retain dapdiamide A synthase. This antibiotic synthase-related gene may contribute to the high stress tolerance documented in *Durusdinium* species, and its increased expression during endosymbiosis suggests increased antibacterial activity within the symbiotic complex.

Keywords: RNA-seq; scleractinian coral; Symbiodiniaceae; endosymbiosis

Citation: Yuyama, I.; Ugawa, N.; Hashimoto, T. Transcriptome Analysis of *Durusdinium* Associated with the Transition from Free-Living to Symbiotic. *Microorganisms* **2021**, *9*, 1560. https://doi.org/10.3390/microorganisms9081560

Academic Editor: Shauna Murray

Received: 14 April 2021
Accepted: 12 July 2021
Published: 22 July 2021

Publisher's Note: MDPI stays neutral with regard to jurisdictional claims in published maps and institutional affiliations.

1. Introduction

The dinoflagellate of the Symbiodiniaceae family live symbiotically with a variety of marine invertebrates, including clams, sea slugs, sea anemones, foraminifera, and corals [1–3]. Among these, the symbiotic relationships between symbiodiniacean algae and cnidarians have been studied extensively. Symbiotic algae provide photosynthetic products to corals and receive nitrogen in exchange [4–6]. Published evidence indicates that the activity of symbiotic Symbiodiniaceae is under the control of the host corals [7–9]. In coral cells, algae are present in host-derived acidified vesicles that have carbon-concentrating mechanisms and activate photosynthetic capacity [7]. A recent transcriptome analysis found that dinoflagellate genes involved in molecular chaperoning as well as sugar and ammonia transportation were suppressed during the establishment of endosymbiosis with *Aiptasia* and coral planula larvae [10,11]. Gene expression analyses of actin, Ca^{2+} ATPase, and H^+ APTase in Symbiodiniaceae also revealed that their expression patterns differed considerably between the non-symbiotic and symbiotic states [12,13]. However, despite these recent advances, many aspects of the changes dinoflagellate undergo during coral endosymbiosis establishment remain unclear. Previous studies have established a model endosymbiosis system consisting of monoclonal alga and juvenile corals, and transcriptome data for these coral–alga complexes have been published, with a focus on coral gene expression [11,14–16]. By contrast, the gene expression levels of dinoflagellate during coral endosymbiosis have not been investigated due to a lack of transcriptome data for the non-symbiotic state. Since the coral–alga model system facilitates the investigation of dinoflagellate gene expression and proliferation processes over time, it is well suited for examining changes in dinoflagellate during endosymbiosis.

In this study, we obtained transcriptome data for cultured *Durusdinium* that were previously used in an infection experiment with juvenile corals [14]. Therefore, in order to comprehensively investigate the changes in Symbiodiniaceae associated with the transition to the symbiotic state, we attempted to detect differentially expressed genes between the free-living and endosymbiotic states. Following the inoculation of juvenile corals with *Durusdinium*, its rate of increase was greater than that of *Cladocopium*, with about 300 *Durusdinium* cells per polyp detected by the 10th day of endosymbiosis, and about 600 cells per polyp detected by the 20th [14]. Here, we identified and functionally annotated differentially expressed genes (DEGs) between the non-symbiotic and symbiotic states, and performed phylogenetic analyses for a part of the DEGs to confirm that the DEGs were derived from *Durusdinium*.

2. Materials and Methods

Symbiodiniaceae strains CCMP 2556 (genus *Durusdinium*) were purchased from the Bigelow Laboratory for Ocean Sciences (West Boothbay Harbor, ME, USA; https://ccmp.bigelow.org/ (accessed on 20 Octorber 2011)). Cultures were grown at 24 °C under a 12 h/12 h light/dark cycle at 80 μmol m^{-2} s^{-1}, in 100 mL of filtered seawater. Further details on RNA isolation and sequencing are provided in supplementary data.

Illumina HiSeq2000 transcriptome data for the symbiotic state of *Durusdinium trenchii* were obtained from the DDBJ Sequence Read Archive (accession nos. DRR119964-119967). Data obtained 10 and 20 days after coral incubation with *D. trenchii* [14], and those obtained in the present study, were used for DEG analysis. The quality checking of filtered reads, mapping, and the detection of DEGs between the non-symbiotic and symbiotic states were performed as described by Yuyama et al., 2018, and the supplementary materials. To confirm the calculated RNA-seq results, we performed bootstrap resampling of the raw read data, with 100 replicates per sample using the isoDE2 package [17], and with 100 replicates ($n = 10$) for each free-living or symbiotic sample (duplicate) in order to examine the expression changes between these states. Since the ANOVA using 100 bootstrap resampling results showed considerable variation among replicates for the data collected at day 10 of endosymbiosis, we eliminated these data from our analysis in order to detect changes related to endosymbiosis. Student's *t*-test using 100 resampling results generated undetectably low *p*-values for most genes, so 10 items of data were randomly selected from the 100 replicates (100 bootstrap resampling replicates) and a *t*-test was performed. Differences in the mean (among the 100 replicates) expression of each gene between the two states were detected using the *t*-test ($p < 0.025$). The *p*-value was set to correspond to the number of DEGs detected in edgeR analysis (q = 0.01). These obtained DEGs were compared with those detected via the *TCC* analysis, and those in common were selected. We also selected genes with an expression change of $\log_2$ (fold change) > 1 between states. The functional annotations of the DEGs are described in the supplementary materials.

Among the DEGs, we focused on dapdiamide A synthetase genes characteristic of *Dursdinium*. A phylogenetic analysis of dapdiamide A synthase was performed using the homologous gene sequences derived from diverse organisms in order to confirm that the genes were derived from *Durusdinium* rather than from bacteria. Further details on the phylogenetic analysis are given in the supplementary materials.

3. Results and Discussion

In this study, we attempted to clarify the gene expression changes taking place in dinoflagellate during coral–alga endosymbiosis establishment. We prepared and sequenced a cDNA library derived from the symbiont culture, which isolated 25,068 contigs containing ORFs. Low reads derived from algae engaged in endosymbiosis with corals and free-living cultured algae were mapped against these contigs, and DEGs between these states were identified. The edgeR analysis identified 8543 DEGs, representing 34% of the candidate alga-derived transcripts. To validate these results, we used bootstrap replicates of RNA-seq data to detect DEGs between the non-symbiotic (n = 2) and symbiotic states (n = 2). Differences in the mean (among 100×2 replicates) expression of each gene between the two states were detected using the Student's *t*-test ($p < 0.025$). The p value was set to correspond to the number (8642) of DEGs detected in the edgeR analysis (q = 0.01). A total of 4587 DEGs were common between both groups. Finally, we selected 1214 genes with $\log_2$ (fold change) > 1 between the free-living and symbiotic states in the edgeR analysis (Figure S1). The top 50 genes showing expression changes due to endosymbiosis with the lowest FDR included ribosomal proteins, heat-shock proteins, and chlorophyll-binding proteins (Figure S1). We also searched for GO molecular function terms that were enriched in these 1214 genes (Figure S2). The most enriched GO terms for these upregulated genes included protein–chromophore linkage and photosynthesis. The upregulated DEGs indicated that algae have enhanced photosynthetic activity during endosymbiosis with corals, which is consistent with previous reports of endosymbiosis in Symbiodiniaceae [7]. Seven processes were related to downregulated DEGs, including microtubule-based processes, mRNA splicing via spliceosome, and protein folding. Genes with decreased expression in microtubule-based processes include genes encoding tubulin, which is a component of flagella. This result may reflect the fact that algae lose their flagella inside corals [18]. Furthermore, a large number of ribosomal and chaperone proteins were detected among the downregulated DEGs, suggesting that some translational and protein-folding functions were inactivated following endosymbiosis establishment. Decreased expression of the chaperone gene has also been reported in the genera *Symbiodinium* and *Cladocopium* [11], and may represent a typical response to coral endosymbiosis establishment. It should be noted, however, that some of the DEGs detected include genes that were altered due to environmental differences between the two states. The time of year when the symbiotic and non-symbiotic states were cultured, as well as changes in the light environment, salinity, and CO_2 in the coral cells, may have affected the genes whose expressions were altered. In order to investigate more specific changes in a symbiotic organism, we must more closely replicate the exact conditions of the culture strain and the symbiotic state.

Among the top 100 DEGs, two genes encoding dapdiamide A synthase (TRINITY_DN38519_c0_g1_i5.p1 (Figure 1) and TRINITY_DN38519_c0_g1_i1.p1) were found to be upregulated. Dapdiamide A synthase adds valine to the carboxylate of fumaramoyl-DAP to form dapdiamide A, an antibiotic, in *Pantoea agglomerans* [19]. Few studies have reported on the antibiotic synthase in Symbiodiniaceae; however, a recent large-scale transcriptome analysis identified dapdiamide A synthase in *Symbiodinium* [20]. Therefore, we performed a phylogenetic analysis to investigate whether the gene encoding dapdiamide A synthase is derived from Symbiodiniacea or from bacteria (Figure 2). One of the genes (TRINITY_DN38519_c0_g1_i5.p1) encoding dapdiamide A synthase was used for a BLASTp query against the NCBI database, and 117 sequences were selected for phylogenetic inference. The distribution of eukaryotic dapdiamide A synthase was restricted to large phylogenetic groups including stramenopiles, haptophytes, and alveolates (Symbiodiniaceae). In the ML tree (Figure 2), most of the eukaryotic sequences formed two separate clades, A and B. Clade A comprises sequences derived from bacillariophytes (stramenopiles), haptophytes, and Symbiodiniaceae. Nine Symbiodiniaceae sequences were monophyletic, with 86% support, and its sister group was shared by *Emiliania huxleyi* (haptophyte) sequences. These branching patterns suggest that the eukaryote–eukaryote lateral gene transfer of dapdiamide A synthase occurred between *Emiliania* and Symbiodiniaceae. In addition, close relationships between two *Symbiodinium* sequences as well as archaean (100%) and *Aureococcus* (pelagophyte) (no support) sequences on the lower part of the tree suggest other types of lateral gene transfer involving Symbiodiniaceae. However, these genes were detected only in the genera *Symbiodinium* and *Durusdinium* of Symbiodiniaceae in this study. *Durusdinium* species exhibit high stress resistance, and have been reported to confer this property to their host corals [21]. Our results show that both dapdiamide A synthase genes from *Durusdinium* were upregulated during endosymbiosis establishment, which may enhance antibacterial action and confer stress tolerance to the host coral.

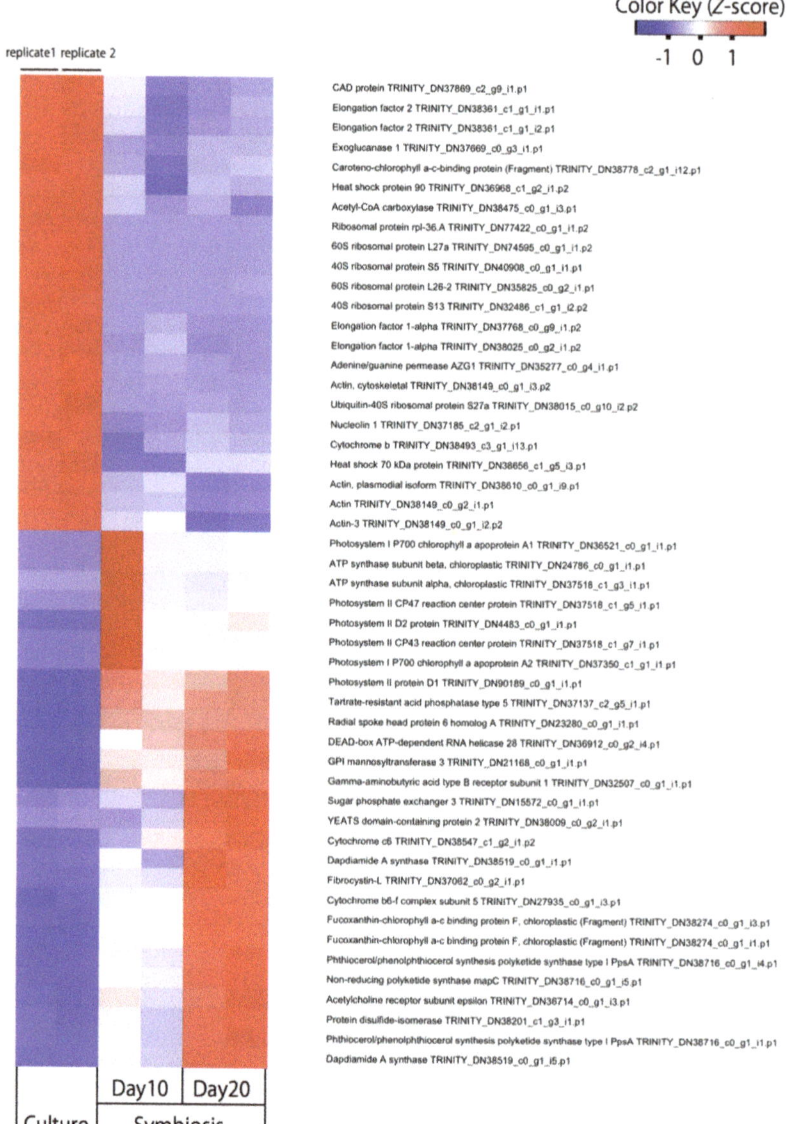

Figure 1. Heat map of RNA-seq analysis for 50 selected genes showing different gene expression pattern in free-living and symbiotic *Durusdinium* at 10 and 20 days post-inoculation (*n* = 2). Among the 1214 DEGs shown in Figure S1, the 50 genes with the lowest false discovery rate are summarized in the heatmap.

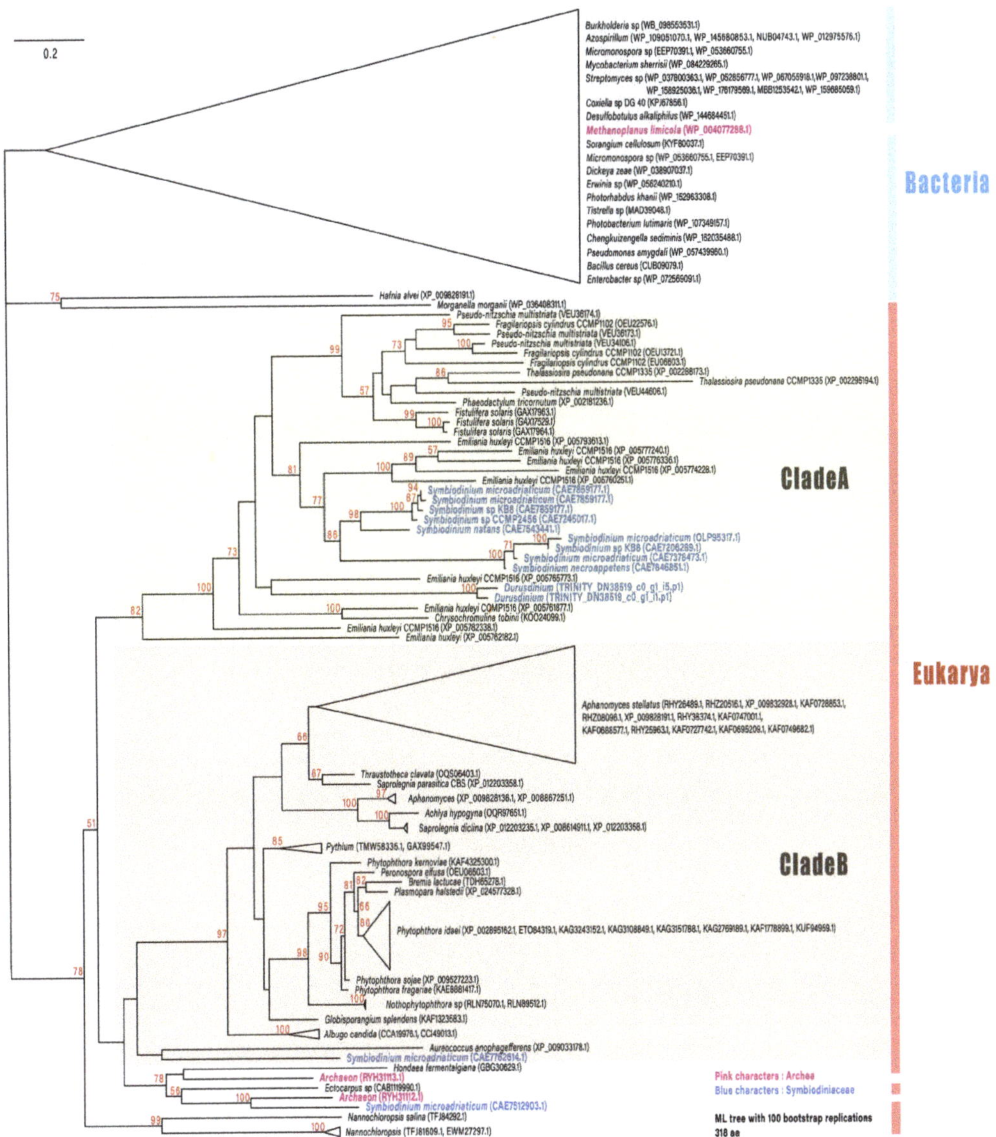

Figure 2. Maximum likelihood phylogenetic tree describing the relationships among dapdiamide A synthase proteins from representative eukaryotes and prokaryotes. All bacterial and one archaean sequence were separated from the remaining eukaryotic/archaeal sequences (78%). Most of the eukaryotic sequences formed two separate clades, A and B. In clade A, bacillariophyte sequences were monophyletic (99%), excluding haptophyte and Symbiodiniaceae sequences. Nine *Symbiodinium* sequences were monophyletic, with 86% support, and its sister group was shared by *Emiliania huxleyi* (haptophyte) sequences. Numbers in red near the nodes are ultrafast bootstrap support values; values < 50% are not included. Blue indicates genes derived from Symbiodiniaceae.

In this study, *Durusdinium* genes that exhibited expression changes during coral endosymbiosis establishment were selected using two analysis methods for functional analysis. The weakness of this study is the small number of replicas and the different timings of the fixation of symbiotic and non-symbiotic dinoflagellate. In future gene expression research, it will be necessary to improve these areas. In addition, transcriptome data do not necessarily correlate with protein expression data, thus requiring proteome analysis to elucidate the entire internal symbiotic process. The roles of these genes in dinoflagellate adaptation to the host coral environment need to be further investigated; such data could be useful in clarifying the evolutionary process of symbiont trait acquisition.

Supplementary Materials: The following are available online at https://www.mdpi.com/article/10.3390/microorganisms9081560/s1, Additional information on methods, Figure S1: Number of differentially expressed genes (DEGs) identified by comparing non-symbiotic and symbiotic *Durusdinium* using edgeR method and Student's *t*-test with bootstrap resampling data, Figure S2: The results of GO (gene ontology) analysis using downregulated 1025 DEG and upregulated 189 DEG against all *Durusdinium* mRNA sequence isolated our studies.

Author Contributions: Conceptualization, I.Y.; methodology, I.Y. and T.H.; validation, N.U., I.Y., and T.H.; formal analysis, N.U.; investigation, N.U. and I.Y.; resources, I.Y.; data curation, N.U. and I.Y.; writing—review and editing, I.Y. and T.H.; visualization, I.Y.; funding acquisition, I.Y. All authors have read and agreed to the published version of the manuscript.

Funding: This work was supported by a research grant from the Japan Society for the Promotion of Science (#19H03026), and the Environment Research and Technology Development Fund (No. 4–1806) from the Ministry of the Environment in Japan, and the Kurita Water and Environment Foundation (18B083).

Institutional Review Board Statement: Not applicable.

Informed Consent Statement: Not applicable.

Data Availability Statement: The raw fastq files for the RNA-seq libraries were deposited at *DDBJ* Sequence Read Archive (DRA) with the accession number of DRA10343.

Acknowledgments: We would like to thank the members of the Laboratory of Molecular Evolution of Microbes in the University of Tsukuba for valuable discussion. RNA-seq analyses were partially performed on the NIG supercomputer at the ROIS National Institute of Genetics.

Conflicts of Interest: The authors declare no conflict of interest. The funders had no role in the design of the study; in the collection, analyses, or interpretation of data; in the writing of the manuscript; or in the decision to publish the results.

References

1. LaJeunesse, T.C.; Parkinson, J.E.; Gabrielson, P.W.; Jeong, H.J.; Reimer, J.D.; Voolstra, C.R.; Santos, S.R. Systematic revision of Symbiodiniaceae highlights the antiquity and diversity of coral endosymbionts. *Curr. Biol.* **2018**, *28*, 2570–2580.e6. [CrossRef] [PubMed]
2. Pochon, X.; Gates, R.D. A New *Symbiodinium* clade (*Dinophyceae*) from soritid foraminifera in Hawai'i. *Mol. Phylogenet. Evol.* **2010**, *56*, 492–497. [CrossRef]
3. Pochon, X.; Montoya-Burgos, J.I.; Stadelmann, B.; Pawlowski, J. Molecular phylogeny, evolutionary rates, and divergence timing of the symbiotic dinoflagellate genus *Symbiodinium*. *Mol. Phylogenet. Evol.* **2006**, *38*, 20–30. [CrossRef]
4. Tanaka, Y.; Miyajima, T.; Koike, I.; Hayashibara, T.; Ogawa, H. Translocation and conservation of organic nitrogen within the coral-zooxanthella symbiotic system of *Acropora pulchra*, as demonstrated by dual isotope-labeling techniques. *J. Exp. Mar. Biol. Ecol.* **2006**, *336*, 110–119. [CrossRef]
5. Whitehead, L.F.; Douglas, A.E. Metabolite comparisons and the identity of nutrients translocated from symbiotic algae to an animal host. *J. Exp. Biol.* **2003**, *206*, 3149–3157. [CrossRef]
6. Wang, J.T.; Douglas, A.E. Essential amino acid synthesis and nitrogen recycling in an alga-invertebrate symbiosis. *Mar. Biol.* **1999**, *135*, 219–222. [CrossRef]
7. Barott, K.L.; Venn, A.A.; Perez, S.O.; Tambutté, S.; Tresguerres, M. Coral host cells acidify symbiotic algal microenvironment to promote photosynthesis. *Proc. Natl. Acad. Sci. USA* **2015**, *112*, 607–612. [CrossRef]
8. Takeuchi, R.; Jimbo, M.; Tanimoto, F.; Iijima, M.; Yamashita, H.; Suzuki, G.; Harii, S.; Nakano, Y.; Yasumoto, K.; Watabe, S. *N*-Acetyl-D-Glucosamine-binding lectin in *Acropora tenuis* attracts specific Symbiodiniaceae cell culture strains. *Mar. Drugs* **2021**, *19*, 146. [CrossRef]
9. Kuniya, N.; Jimbo, M.; Tanimoto, F.; Yamashita, H.; Koike, K.; Harii, S.; Nakano, Y.; Iwao, K.; Yasumoto, K.; Watabe, S. Possible involvement of Tachylectin-2-like lectin from *Acropora tenuis* in the process of *Symbiodinium* Acquisition. *Fish. Sci.* **2015**, *81*, 473–483. [CrossRef]

10. Maor-Landaw, K.; van Oppen, M.J.H.; McFadden, G.I. Symbiotic lifestyle triggers drastic changes in the gene expression of the algal endosymbiont *Breviolum minutum* (Symbiodiniaceae). *Ecol. Evol.* **2020**, *10*, 451–466. [CrossRef] [PubMed]

11. Mohamed, A.R.; Andrade, N.; Moya, A.; Chan, C.X.; Negri, A.P.; Bourne, D.G.; Ying, H.; Ball, E.E.; Miller, D.J. Dual RNA-sequencing analyses of a coral and its native symbiont during the establishment of symbiosis. *Mol. Ecol.* **2020**, *29*, 3921–3937. [CrossRef]

12. Bertucci, A.; Tambutté, É.; Tambutté, S.; Allemand, D.; Zoccola, D. Symbiosis-dependent gene expression in coral–dinoflagellate association: Cloning and characterization of a P-Type H+-ATPase gene. *Proc. R. Soc. B Biol. Sci.* **2010**, *277*, 87–95. [CrossRef] [PubMed]

13. Watanabe, T.; Kii, S.; Tanaka, J.; Takishita, K.; Maruyama, T. cDNA cloning and phylogenetic and expression analyses of actin in symbiotic dinoflagellates (*Symbiodinium* spp.). *J. Appl. Phycol.* **2006**, *18*, 219–225. [CrossRef]

14. Yuyama, I.; Ishikawa, M.; Nozawa, M.; Yoshida, M.; Ikeo, K. Transcriptomic changes with increasing algal symbiont reveal the detailed process underlying establishment of coral-algal symbiosis. *Sci. Rep.* **2018**, *8*, 16802. [CrossRef]

15. Yuyama, I.; Higuchi, T. Comparing the effects of symbiotic algae (*Symbiodinium*) clades C1 and D on Early growth stages of *Acropora tenuis*. *PLoS ONE* **2014**, *9*, e98999. [CrossRef] [PubMed]

16. Yoshioka, Y.; Yamashita, H.; Suzuki, G.; Zayasu, Y.; Tada, I.; Kanda, M.; Satoh, N.; Shoguchi, E.; Shinzato, C. Whole-genome transcriptome analyses of native symbionts reveal host coral genomic novelties for establishing coral–algae symbioses. *Genome Biol. Evol.* **2021**, *13*, evaa240. [CrossRef]

17. Al Seesi, S.; Tiagueu, Y.T.; Zelikovsky, A.; Măndoiu, I.I. Bootstrap-based differential gene expression analysis for RNA-Seq data with and without replicates. *BMC Genom.* **2014**, *15*, S2. [CrossRef] [PubMed]

18. Camaya, A.P. Stages of the symbiotic zooxanthellae–host cell division and the dynamic role of coral nucleus in the partitioning process: A novel observation elucidated by electron microscopy. *Coral Reefs* **2020**, *39*, 929–938. [CrossRef]

19. Hollenhorst, M.A.; Clardy, J.; Walsh, C.T. The ATP-dependent amide aigases DdaG and DdaF assemble the fumaramoyl-dipeptide scaffold of the Dapdiamide antibiotics. *Biochemistry* **2009**, *48*, 10467–10472. [CrossRef]

20. Silva Lima, A.W.; Leomil, L.; Oliveira, L.; Varasteh, T.; Thompson, J.R.; Medina, M.; Thompson, C.C.; Thompson, F.L. Insights on the genetic repertoire of the coral *Mussismilia braziliensis* endosymbiont *Symbiodinium*. *Symbiosis* **2020**, *80*, 183–193. [CrossRef]

21. Stat, M.; Gates, R.D. Clade D *Symbiodinium* in Scleractinian Corals: A "Nugget" of Hope, a Selfish Opportunist, an Ominous Sign, or All of the Above? *J. Mar. Biol.* **2010**, *2011*, e730715. [CrossRef]

 *microorganisms*

Article

sxtA4+ and sxtA4- Genotypes Occur Together within Natural *Pyrodinium bahamense* Sub-Populations from the Western Atlantic

Kathleen Cusick * and Gabriel Duran

Department of Biological Sciences, University of Maryland Baltimore County, Baltimore, MD 21250, USA; gaduran1@umbc.edu
* Correspondence: kcusick@umbc.edu

Abstract: Saxitoxin (STX) is a secondary metabolite and potent neurotoxin produced by several genera of harmful algal bloom (HAB) marine dinoflagellates. The basis for variability in STX production within natural bloom populations is undefined as both toxic and non-toxic strains (of the same species) have been isolated from the same geographic locations. *Pyrodinium bahamense* is a STX-producing bioluminescent dinoflagellate that blooms along the east coast of Florida as well as the bioluminescent bays in Puerto Rico (PR), though no toxicity reports exist for PR populations. The core genes in the dinoflagellate STX biosynthetic pathway have been identified, and the *sxtA4* gene is essential for toxin production. Using *sxtA4* as a molecular proxy for the genetic capacity of STX production, we examined *sxtA4+* and *sxtA4-* genotype frequency at the single cell level in *P. bahamense* populations from different locations in the Indian River Lagoon (IRL), FL, and Mosquito Bay (MB), a bioluminescent bay in PR. Multiplex PCR was performed on individual cells with *Pyrodinium*-specific primers targeting the 18S rRNA gene and *sxtA4*. The results reveal that within discrete natural populations of *P. bahamense*, both *sxtA4+* and *sxtA4-* genotypes occur, and the *sxtA4+* genotype dominates. In the IRL, the frequency of the *sxtA4+* genotype ranged from ca. 80–100%. In MB, *sxtA4+* genotype frequency ranged from ca 40–66%. To assess the extent of *sxtA4* variation within individual cells, *sxtA4* amplicons from single cells representative of the different sampling sites were cloned and sequenced. Overall, two variants were consistently obtained, one of which is likely a pseudogene based on alignment with cDNA sequences. These are the first data demonstrating the existence of both genotypes in natural *P. bahamense* sub-populations, as well as *sxtA4* presence in *P. bahamense* from PR. These results provide insights on underlying genetic factors influencing the potential for toxin variability among natural sub-populations of HAB species and highlight the need to study the genetic diversity within HAB sub-populations at a fine level in order to identify the molecular mechanisms driving HAB evolution.

Keywords: bioluminescence; dinoflagellate; *Pyrodinium bahamense*; harmful algal blooms; saxitoxin; *sxtA4*

Citation: Cusick, K.; Duran, G. sxtA4+ and sxtA4- Genotypes Occur Together within Natural *Pyrodinium bahamense* Sub-Populations from the Western Atlantic. *Microorganisms* **2021**, *9*, 1128. https://doi.org/10.3390/microorganisms9061128

Academic Editor: Shauna Murray

Received: 1 April 2021
Accepted: 11 May 2021
Published: 23 May 2021

1. Introduction

Saxitoxin is a secondary metabolite produced by several genera of dinoflagellates and cyanobacteria that is better known as a potent neurotoxin due to its detrimental effects on human health. It is the parent molecule in a class of compounds collectively referred to as paralytic shellfish toxins (PSTs), which target voltage-gated ion channels (sodium, potassium, calcium) in humans and can lead to death via respiratory paralysis (as reviewed in [1]). It causes the human illnesses paralytic shellfish poisoning (PSP) and saxitoxin pufferfish poisoning (SPFP) [2], in which shellfish and pufferfish, respectively, ingest the toxic cells and bioaccumulate the toxins within their organs and tissues. STX is produced by microbes from two kingdoms of life inhabiting different aquatic habitats: dinoflagellates (Eukaryota) in marine systems [3–6], and cyanobacteria (Bacteria) in freshwater systems [7,8], though the biosynthetic pathway is similar between the two groups [9,10].

The gene cluster for STX biosynthesis was first identified in the toxic cyanobacterium *Cylindrospermopsis raciborskii* T3 [11] followed by identification of homologous gene clusters in other cyanobacteria [12,13]. The core genes in STX biosynthesis have more recently been identified in the three marine dinoflagellate genera capable of STX production-multiple *Alexandrium* spp., *Pyrodinium bahamense*, and *Gymnodinium catenatum*–albeit with varying degrees of coverage [14,15]. The first gene in the pathway, *sxtA*, [14], codes for a novel polyketide synthase comprised of four catalytic domains (SxtA1, SxtA2, SxtA3, and SxtA4) [16]. Two different forms of SxtA occur in dinoflagellates: a long transcript yielding all four domains (coded for by the genes *sxtA1, 2, 3, and 4*), and a short transcript lacking *sxtA4* [14]. The gene located at the N-terminal portion of *sxtA* (*sxtA1*), coding for the acyltransferase and phosphopantetheinyl-attachment site domain, is widespread, occurring in both toxic and non-toxic strains of toxic species as well as non-toxic species from other genera [15,17,18]. *SxtA4*, which codes for an amidotransferase [11] appears to be relatively specific for toxin synthesis, as demonstrated by its presence and sequence conservation among numerous toxic strains encompassing all three STX-producing genera and its absence in non-STX-producing species [14,19–21].

Multiple examples exist as to both the phenotypic and genetic diversity occurring within monospecific HAB populations and/or blooms. Broad intra-specific variability in multiple traits including toxin production, growth rate, and motility, have been documented among strains (of the same species) isolated from the same water sample or geographic area for various HAB species [22–25]. In general, a dinoflagellate strain is obtained by establishing a clonal culture, which is derived from the isolation of a single cell from an environmental water sample [24]. Clonal diversity has been shown to be linked to differences in genome size [26], morphology [27], and growth rate [28]. An increase in the use of molecular tools in dinoflagellate population ecology studies have revealed that dinoflagellates are not monoclonal, even in a monospecific bloom [24,29]. High levels of genetic diversity are well-documented among toxic *Alexandrium* dinoflagellate species from the same geographic area, primarily through assessment with microsatellite markers [25,29].

Numerous studies have documented toxin variability in strains isolated from the same geographic area for all three STX-producing genera [22,30,31]. For example, analysis of multiple *A. minutum* strains from the coast of Ireland found strains from southern areas to be toxic but not those from the west coast [22]. Natural populations of *A. tamarense* have been shown to be comprised of a mix of strains, differing in both toxin content and profiles [23]. Studies with a large range of *Gymnodinium* strains demonstrated a high level of intra-population and regional variation in both the presence and amounts of STX congeners and profiles [30]. Additionally, while *P. bahamense* toxic outbreaks are common in Mexico, screening numerous strains from the same area found only one isolate to be a confirmed toxin-producer [31].

The collective findings with microsatellite markers that monospecific dinoflagellate blooms are not monoclonal events coupled with the isolation of both toxic and non-toxic strains of the same species from similar geographic regions underscores the need to examine the genetic capacity for toxin production as a contributor to toxin variability within bloom sub-populations. In the case of the *A. minutum* strains from the south (toxic) and west (non-toxic) coasts of Ireland, the strain differentiation–and by extension toxin potential—was not possible based on common phylogenetic markers, with toxin confirmation achieved only via HPLC analysis [22]. Population genetic studies with *Alexandrium* showed that genetically similar populations (defined via microsatellite markers) differ in toxin production both within and across populations, with natural populations comprised of a mix of strains that differ in both toxin content and profiles [23,25]. Variation in the *sxtA4* genotype in natural populations of dinoflagellate species capable of STX production remains unknown.

P. bahamense is a bioluminescent dinoflagellate found in both the Atlantic-Caribbean and the Indo-Pacific [32,33]; toxic outbreaks are well-documented from *P. bahamense* in the

Indo-Pacific [5,34–38]. In contrast, *P. bahamense* bloomed in the Indian River Lagoon (IRL), along the east coast of Florida, for years with no known record of STX production. Blooms of *P. bahamense* are the source of the bioluminescence found along both the Florida coast and in the bioluminescent bays in Puerto Rico. However, in the mid-2000s, contaminated seafood outbreaks from multiple states in the US were traced back to Florida, and *P. bahamense* was identified as the source [2,39], marking the first occurrence of toxin production in the Western Atlantic. No genetic data were collected for these sub-populations, and so the molecular mechanisms (*sxt* gene presence and/or regulation) underlying this toxin production remain unknown.

A large gap in our understanding is the factors influencing toxin variability among natural dinoflagellate bloom populations, as both toxic and non-toxic strains of the same species have been isolated from the same geographic locations [31,40]. *SxtA4* is essential for STX biosynthesis [11,14,17,41,42]. Other dinoflagellate genes have been shown to undergo independent gene duplication and shuffling processes [43,44], and in recent years horizontal gene transfer has been demonstrated to be a significant driver in dinoflagellate genome evolution [45,46]. Analysis based primarily on *Alexandrium* isolates indicates that *sxtA4* is lost rather than gained, and that widespread horizontal gene transfer does not occur for this gene [17].

While the core genes in the dinoflagellate STX biosynthetic pathway are known [14,15], how this relates to variability in STX production within natural populations from the same geographic area remains unknown. Recent studies found *sxtA4* gene presence but not transcription in several non-toxic strains of toxic *Alexandrium* spp. [14,17,19,42]. Thus, while "non-toxic" strains may or may not possess the *sxtA4* gene, lack of *sxtA4* is a strong indication of lack of toxicity. In this study, we used *sxtA4* presence within the genome of individual cells as a molecular proxy as to the genetic potential for toxin production among *P. bahamense* cells in a natural population. We use the term "sxtA4+ genotype" to indicate cells in which the sxtA4 gene was detected in the genome; conversely, "sxtA4-genotype" indicates sxtA4 was not found. We examined *sxtA4+* genotype frequency (the proportion of the *sxtA4+* genotype within a population) and *sxtA4* gene variation among single cells within natural *P. bahamense* sub-populations from multiple locations in Florida and Puerto Rico. Our data reveal that both sxtA4+ and sxtA4- genotypes occur within the same sub-population and that frequency of the *sxtA4+* genotype dominates in Florida populations.

2. Materials and Methods

2.1. sxtA4 Genomic Characterization from P. bahamense Lab Isolate

Genomic DNA and total RNA were extracted from 15 mL (1.5×10^5 cells) of a toxin-producing *P. bahamense* isolate from the IRL using the TRI Reagent kit (Invitrogen, Carlsbad, CA, USA) following manufacturer's protocol. DNA was diluted 1:10 for subsequent PCRs. RNA was reverse transcribed using the High Capacity RNA-to-cDNA kit (ABI, Waltham, MA, USA), using 1 ug of total RNA. Reactions were performed in a Veriti thermocycler (ThermoFisher, Waltham, MA, USA) under the following conditions: 37 °C for 60 min, followed by 95 °C for 5 min. CDNA was then diluted 1:10 for subsequent PCRs. *SxtA4* DNA was initially amplified using the primers sxt007 and sxt008 (Table 1) with the conditions described by Stuken [14], which yielded a 750 bp product. Additionally, primers were designed based on all publicly-available *P. bahamense sxtA4* sequences in GenBank as of 13 August 2019 to target as nearly a full-length *Pyrodinium*-specific *sxtA4* as possible. This yielded the primer pair sxtA4F1/sxtA4680R (Table 1), which amplified an ca. 680 bp region. PCRs were performed with the ThermoFisher Taq polymerase PCR kit (Waltham, MA, USA) as follows (as final concentrations): $1 \times$ KCl (-Mg) buffer, 1.5 mM $MgCl_2$, 200 nM each dNTP, 400 nM each forward and reverse primer, 0.625 U Taq DNA polymerase, 2 μL DNA diluted 1:10, and brought to a final volume of 25 μL with nuclease-free water. A touchdown PCR was used that consisted of an initial denaturation at 95 °C for 2 min followed by 2 cycles of 95 °C, 30 s, 58 °C, 15 s, 72 °C, 45 s, and subsequent two cycles with annealing temperatures of 54 °C, 52 °C, 50 °C, 50 °C, 46 °C, and 35 cycles at 44 °C, with a final extension at 72 °C for 7 min. The primer pair sxt007/sxtA4680R was then used to amplify an 815 bp region from both genomic DNA and cDNA. PCRs and thermocycling conditions were as described for the sxtA4F1/680R PCRs. For all PCRs, the resulting amplicons were cloned into the pCR4 vector using the TOPO TA cloning kit (Invitrogen, Carlsbad, CA, USA) and transformed into TOP10 chemically competent *Escherichia coli* cells following the manufacturer's instructions. Transformed cells were plated onto Miller's Luria Broth (LB) plates containing 50 mg mL^{-1} kanamycin (LB$_{kan50}$) and incubated overnight at 37 °C. The resulting transformants were screened via colony PCR using the M13 forward (F) and M13 reverse (R) primers followed by gel electrophoresis to identify those with inserts of the expected size. Colony PCRs were conducted with the ThermoFisher Taq polymerase PCR kit as follows (final concentrations): 1× KCl (-Mg) buffer, 1.5 mM MgCl$_2$, 200 nM each dNTP, 200 nM each M13F and M13R primers, 1 U Taq DNA polymerase, and brought to a final volume of 20 uL with nuclease-free water. Thermocycling conditions consisted of an initial denaturation/cell lysis of 95 °C, 3 min, followed by 35 cycles of 95 °C, 15 s, 50 °C, 15 s, 72 °C, 1 min, and a final extension of 72 °C for 7 min. Colonies producing a band of the expected size were inoculated into 2 mL LB$_{kan50}$ broth and grown overnight at 37 °C. The plasmids were isolated using the QIAprep Spin Miniprep kit (Qiagen, Hilden, Germany) and sequenced in both directions using the M13 F and R primers by GeneWiz, Inc. (Germantown, MD, USA). Sequence data were trimmed of primer and vector sequences and evaluated using the BLAST program [47] with comparison against published sequences in GenBank. Sequences from multiple clones were aligned in MEGA7 using ClustalW and manually adjusted.

Table 1. List of primers.

Primer	Sequence	Source
PyrosxtA4F1	AACGACATGAAGCAGCTCGA	this study
PyrosxtA4R680	CTAGATGGGGTACCACATAG	this study
SxtA4166F	CATGGCTGCGGCGTTCTTG	this study
Pcomp370F	AAATTACCCAATCCTGACACT	48
Pcomp1530R	CTGATGACTCAGGCTTACT	48
sxt007	ATGCTCAACATGGGAGTCATCC	14
sxt008	GGGTCCAGTAGATGTTGACGATG	14

The 815 bp *stxA4* DNA sequence was aligned with existing *sxtA4* sequences for *Alexandrium* and *Gymnodinium* and used to design additional forward and reverse primers within *sxtA4*. The primer pair sxtA4F1/680R yielded the longest *Pyrodinium*-specific *sxtA4* sequence and so was then optimized for use in a single-cell multiplex PCR assay with *P. bahamense*-specific 18S rRNA gene primers. The sxtA4166F primer targeted a conserved region within *sxtA4* and so was used in subsequent nested PCRs and sequencing reactions (described below).

2.2. sxtA4 Genotype Analysis from Single Cells

2.2.1. Sample Collection

Whole water samples were collected from multiple locations in the IRL and Mosquito Bay, PR, from July–September 2019 and 2020 from the surface to a depth of 0.5 m (Figure 1). Samples from the IRL were collected onto 30-micron mesh and the biomass collected on the mesh rinsed into a sterile 1 L bottle using filtered water. In Mosquito Bay, whole-water grab samples were collected without size fractionation. Samples were transported at ambient temperature back to the laboratory for processing.

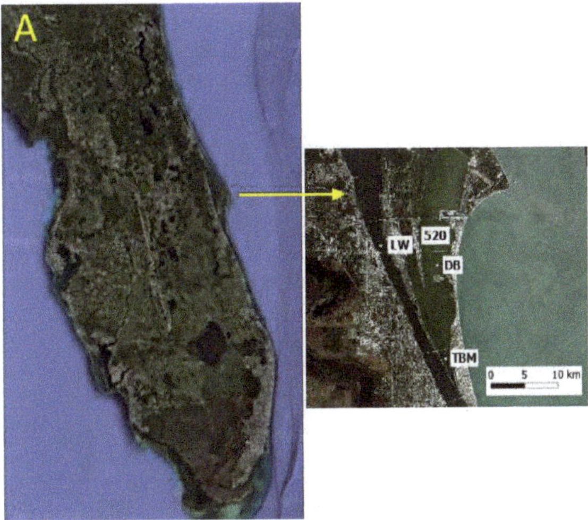

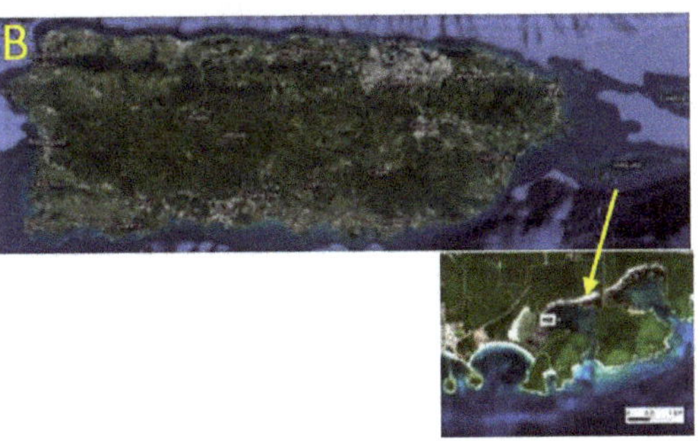

Figure 1. Map depicting sampling sites in (**A**) Indian River Lagoon, FL and (**B**) Mosquito Bay in Vieques, Puerto Rico. Florida sampling sites: TBM = Telemar Bay Marina, DB = Diamond Bay, LW = Lee Wenner Park, 520 = 520 Bridge. MB (Vieques map) = Mosquito Bay.

2.2.2. Cell Isolation and Lysis

Single cells were isolated for lysis and subsequent multiplex PCR using a slight modification of a protocol described previously [48]. Briefly, an aliquot (ca. 1–3 mL) of the environmental sample water was placed in a petri dish and viewed with an inverted light microscope. *P. bahamense* cells were identified under 400× magnification based on morphology [49]. At all sampling sites, *P. bahamense* was the dominant dinoflagellate species. Individual cells were isolated with a sterile glass micropipette under 100× magnification, washed twice in sterile HPLC water, and placed in sterile 200 µL PCR tubes. Samples were stored at −20 °C until DNA extraction. Prior to DNA extraction, the final volume of water in each tube was brought to 20 uL with nuclease-free water. DNA extraction consisted of five consecutive freeze-thaw cycles alternating between baths of a dry ice/ethanol slurry and heating to 100 °C. Tubes were centrifuged briefly prior to PCR.

2.2.3. Multiplex PCR

Multiplex PCR was performed on single cells using *Pyrodinium*-specific primers targeting the 18S rRNA gene developed previously [48] and the sxtA4F1/stxA4680R primers. The single-cell multiplex PCR was first optimized on single cells from lab cultures of the toxin-producing *P. bahamense* isolate from the IRL using the same methods for cell isolation and washing, lysis, and PCR as for environmental samples. While all reagents of the PCR were optimized, those which significantly improved the reaction included an increase in final magnesium (2.5 mM) and dNTP (0.3 mM) concentrations as well as a decreased concentration of the 18S rRNA gene primers (200 nM final concentration) in comparison to the sxtA4 F1/680R primers (400 nM final concentration). All single cells from the toxic lab isolate that yielded an 18s rRNA gene amplicon also yielded the 680 bp *sxtA4* amplicon (data not shown), indicating the ability of both primer sets to bind to and amplify their respective targets when used in the same PCR. Following the optimization, PCRs were performed on single cells from environmental samples. Single cell multiplex PCRs utilized the GoTaq PCR Core System I kit (Promega, Madison, WI, USA) and consisted of the following (as final concentrations): $1\times$ Colorless GoTaq Flexi buffer, 2.5 mM $MgCl_2$, 0.3 mM dNTP mix, 200 nM each Pcomp370F and Pcomp1530R, 400 nM each sxtA4F1 and sxtA4680R, 3.7 U GoTaq DNA polymerase, and brought to a final volume of 50 μL with 30 μL nuclease-free water. Thermocycling conditions consisted of an initial denaturation at 95 °C for 3 min followed by 45 cycles of 95 °C for 30 s; 57 °C for 15 s; 72 °C for 1 min; and a final extension at 72 °C for 7 min. Multiplex PCRs were also performed on a subset of samples (ca. 40 single cells) with the *sxtA4* primer pair sxt007/sxt008, as this primer pair has been used in all *sxtA4* analyses to date for *Alexandrium*, *Gymnodinium*, and *Pyrodinium*, and thus appears to be conserved across species. Products were visualized on 1% E-Gel EX agarose gels with the E-Gel Power Snap Electrophoresis System (Invitrogen). The 18S rRNA gene served as a positive control to confirm the presence of the cell in the tube and that reactions were not inhibited by potential contaminants. Only samples yielding the 18S rRNA gene amplicon were included in *sxtA4* genotype frequency analysis.

Samples yielding products indicative of both the 18S rRNA gene (1200 bp) and *sxtA4* (680 bp) were further screened by one (or more) additional means to confirm *sxtA4* amplicon specificity. These included: (1) the multiplex PCRs were PCR purified using the Wizard S/V Gel and PCR Clean-Up System (Promega, Madison, WI) and sequenced (as two separate samples) with the internal primer sxtA4166F and the Pcomp350F primer. The sequencing protocol was modified so that the amount required for each product was doubled; (2) bands corresponding to the 18S rRNA gene and *sxtA4* amplicons were gel-purified using the Wizard S/V Gel and PCR Clean-Up System and sequenced with the Pcomp350 F and sxtA4166F primers, respectively; (3) a nested PCR was performed in which the original multiplex PCR product was diluted 1:5 and 2 uL used as the template for a PCR using the sxtA4166F internal primer and 680R. These PCRs utilized the GoTaq PCR Core System I and consisted of the following (as final concentrations): $1\times$ Colorless GoTaq Flexi buffer, 1.5 mM $MgCl_2$, 0.1 mM dNTP mix, 400 nM each sxtA166F and sxtA4680R, 0.625 U GoTaq DNA polymerase, and brought to a final volume of 25 μL with nuclease-free water. Thermocycling conditions consisted of 95 °C for 3 min, followed by 35 cycles of 95 °C, 30 s, 57 °C, 15 sec, 72 °C, 40 s, and a final extension at 72 °C for 7 min. The resulting product was visualized with gel electrophoresis; with this primer set, an amplicon of ca. 500 bp was indicative of *sxtA4*. These samples were then PCR purified and sequenced with 166F to confirm *sxtA4* specificity.

Samples in which only a band indicative of the 18S rRNA gene was amplified were PCR purified and sequenced with Pcomp350F to confirm *P. bahamense* specificity. For this, 20 uL of the initial multiplex was cleaned using the Wizard PCR Clean-Up kit. Subsequent PCRs were then performed on the initial multiplex with the primer sets 166F/680R and F1/680R to explore whether variants were missed using the *Pyrodinium*-specific sxtA4F1 (see details in Supplementary Materials). As the initial multiplex PCR was performed in the same tube that contained the single cell, the genomic DNA, and thus template, was still present in the tube for these PCRs.

To confirm their identity, the obtained partial sequences of *sxtA4* and the 18S rRNA gene were queried against the GenBank nr database using standard BLASTN 2.2.26+ [50].

To examine the extent of *sxtA4* gene variation within individual cells from different populations, the *sxtA4* amplicon (yielded with the 166F/680R primer pair for an amplicon length of ca. 500 bp) from a small subset of individual cells (one representative of each site in the IRL plus MB) was cloned, and a minimum of 10 clones from each sequenced. Amplicons were cloned into the pCR-4 vector using the TOPO TA cloning kit (Invitrogen), transformed into chemically competent TOP10 *E. coli* cells following the manufacturer's instructions for kanamycin selection, and sequenced using the M13 forward and reverse primers by GeneWiz (Germantown, MD, USA).

Sequence data were trimmed of primer and vector sequences and initially queried against the GenBank nr database using standard BLASTN 2.2.26+ [50]. All subsequent analysis was performed using MEGA (Molecular Evolutionary Genetics Analysis) v7. The phylogenetic relationship among sequences in relation to geographic location (as defined by the four sampling sites) was examined to assess whether sequences were specific for a location. *SxtA4* clone sequences obtained from individual cells representative of the four sampling sites as well as *Alexandrium* spp. *sxtA4* sequences retrieved from GenBank were used in the analysis. The model that best fit the data was the Tamura 3-parameter (T92) and so was used to generate Neighbor Joining and minimal evolution trees. Both trees produced identical results. The sequence data were also compared to existing *Pyrodinium sxtA4* sequences in GenBank (MN431957.1, a 358bp fragment, designated as var. *compressum* from Sepanggar Bay, Malaysia; and a 681 bp sequence, from TSA GBXF01000001.1, also designated as var. *compressum*).

Sequences obtained in this study have been have been deposited into GenBank with the following accession numbers: MZ234664-MZ234695.

3. Results and Discussion

3.1. SxtA4 Genomic Characterization

The ca 815 bp *sxtA4* sequence amplified here is the longest to date for *P. bahamense*. Genomic DNA sequences were highly similar among clones, ranging from 96–100%. This was also the case for the cDNA clones, albeit with even greater similarity (99–100%) and overall, the two types of sequences showed a high degree of similarity (Table 2). Alignment of sequences amplified from genomic DNA demonstrated minimal sequence diversity, with 29 base substitutions. Alignment of sequences derived from cDNA showed even less diversity, with only seven base substitutions. This is in keeping with the trend recorded for *Alexandrium*, in that the overall percent similarity was higher for cDNA copies than gDNA copies, indicating that only certain copies are transcribed [14,19]. Phylogenetic analysis demonstrated *sxtA4* for *P. bahamense* from the IRL clustered with *sxtA4* from a toxic Indo-Pacific strain, and these sequences grouped with *Alexandrium*, with *Gymnodinium* forming a separate cluster (Figure 2).

Table 2. Percent similarity between genomic and cDNA *sxtA4* sequences for *P. bahamense*.

Species	No. cDNA Sequences	% Similarity	No. DNA Sequences	% Similarity
P. bahamense IRL	8	99.1–100	14	96.0–100
A. fundyense [1] CCMP1719	12	96.2–99.4	10	91.7–100
A. fundyense A8 [1]	7	93.5–100	19	77.5–100
A. fundyense E4 [1]	19	97.0–100	40	76.8–100
A. pacificum ACCC01 [1]	4	96.0–100	5	91.0–100

[1] *Alexandrium* data [14,17] are provided for comparison.

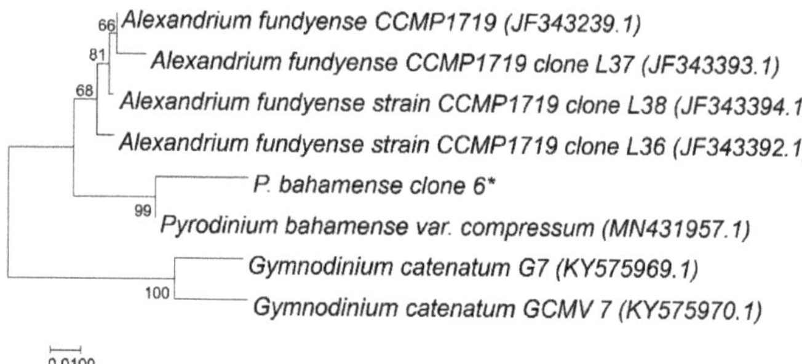

Figure 2. Neighbor Joining tree of *sxtA4* sequences. Representative *Alexandrium*, *Pyrodinium*, and *Gymnodinium sxtA4* sequences were downloaded from GenBank and aligned and trimmed in MEGA 7. Percentages from the bootstrap test (500 replicates) are displayed next to the branches. The rate variation among sites was modeled with a gamma distribution (shape parameter = 0.32). The analysis encompassed 8 nucleotide sequences. All positions containing gaps and missing data were eliminated. There were a total of 575 positions in the final dataset. Evolutionary analyses were conducted in MEGA7. * indicates *P. bahamense sxtA4* sequence from this study.

In comparing the gDNA sequences of the *P. bahamense* lab isolate, two main variants occurred, as defined by consistent base substitutions at 29 sites (i.e., base 182, 4 clones possess "G", the other 12 clones a "C"; base 234, the same 4 clones possess "T", the other 12 a "C"). This was in addition to 27 sites in which the base of an individual clone differed from the others. It has been suggested that harboring multiple, slightly different genomic copies of *sxtA4* may be the underlying basis for the diversity in STX congener profiles and toxicity within *Alexandrium*. The *P. bahamense* lab isolate consistently yields a STX profile of two congeners: STX and GTX-5 (Table S1), even when grown under various conditions.

3.2. sxtA4+/sxtA4- Genotypes in Natural P. bahamense Sub-Populations

Multiplex PCR screening of individual cells revealed that both *sxtA4+* and *sxtA4-* genotypes exist among natural *P. bahamense* populations, with the *sxtA4+* genotype defined by the amplification of *sxtA4* in individual cells. The data to date show that among all sites, the *sxtA4+* genotype dominates, regardless of sample size (Table 3). However, it is worth noting that sequence data were derived from summer (June–September) samplings. In the IRL, *P. bahamense* blooms during this period. *P. bahamense* is present as both bloom and non-bloom events in the IRL. The northern region has particularly long water residence times, and is prone to intense blooms [51–53]. Major blooms of *P. bahamense* in the IRL are defined as > 100 cells/mL; a several-year study on *P. bahamense* distribution in the IRL recorded densities up to 776 cells/mL [54]. However, since that time, greater cell densities, as well as their locations within the northern IRL, have been observed (Cusick pers. observ.) Microscopic observations of water samples collected at the various sites in the IRL showed *P. bahamense* to be the dominant cell type at all. In the bioluminescent bay in Puerto Rico, *P. bahamense* was also the dominant cell type. While not as well-studied as the IRL, a recent one-year study in Mosquito Bay recorded an overall average of ca. 27 cells mL^{-1}, with abundance ranging from 0–244 cells mL^{-1} [55]. A primary difference among sites is that *P. bahamense* is present in Mosquito Bay year-round, while it (typically) disappears over the fall and winter months in the IRL.

Table 3. *sxtA4+* genotype frequencies from sites in Florida (IRL) and Puerto Rico. [1] indicates number of isolated cells yielding an 18S rRNA gene amplicon. [2] Number of isolated cells yielding the 18S rRNA gene amplicon in which *sxtA4* was detected.

Sampling Site	Date	18S [1]	*sxtA4* [2]	*sxtA4+* Frequency
Diamond Bay (IRL)	7/15/2019	10	8	80%
TBM (IRL)	6/15/2020	64	54	84%
TBM (IRL)	7/20/2020	44	36	82%
Lee Wenner (IRL)	7/20/2020	30	30	100%
520 Bridge (IRL)	7/20/2020	37	37	100%
Mosquito Bay (PR)	7/8/2020	9	6	67%
Mosquito Bay (PR)	9/15/2020	7	3	43%

In the IRL, the frequency of the *sxtA4+* genotype ranged from ca. 80–100% (Table 3, Figure 3). Sampling at Diamond Bay (July 2019) showed the *sxtA4+* genotype to occur at 80% frequency. Comparable numbers were obtained for samplings at Telemar Bay Marina in 2020 in both June and July, with the *sxtA4+* genotype occurring at frequencies of 81% and 84%, respectively. The *sxtA4+* genotype occurred at a frequency of 100% in populations from sites LW and 520. The *sxtA4+* genotype occurred at a much lower frequency in Mosquito Bay (Puerto Rico) populations, at ca 67% in July and decreasing to ca 40% in Sept (Figure 3). While these data must be interpreted with caution due to the low sample size, these are the first data demonstrating the presence of *sxtA4* in *P. bahamense* populations in bioluminescent bays in Puerto Rico.

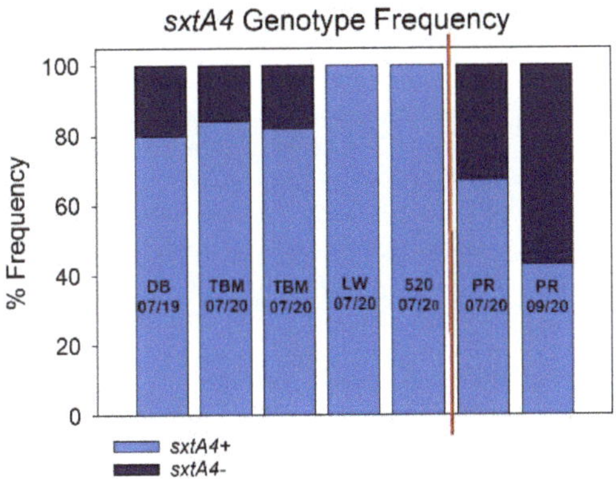

Figure 3. Percent *sxtA4* genotype frequency among sites. Multiplex PCR was performed on individual *P. bahamense* cells collected from multiple sites in the Indian River Lagoon (IRL, Florida) and Mosquito Bay, a bioluminescent bay in Puerto Rico. Cells yielding the *sxtA4* amplicon were defined as "sxtA4+"; cells in which *sxtA4* was not detected were defined as "sxtA4-". Text and numbers in each bar represent sampling site and date: DB = Diamond Bay (IRL); TBM = Telemar Bay Marina (IRL); LW = Lee Wenner Park (IRL); 520 = 520 Bridge (IRL). Vertical red line in graph separates Florida and Puerto Rico (PR) sampling sites.

The correlation between *sxtA4* genomic presence and active toxin production within both Florida and Puerto Rico sub-populations is not known. *P. bahamense* is part of a routine state HAB monitoring program (Florida Fish and Wildlife Conservation Commission) in Florida. Following the initial determination (2002–2004) of *P. bahamense* as the source of saxitoxin in pufferfish and shellfish harvested from the IRL, state agencies established the PSP Biotoxin Contingency Program to monitor saxitoxin in shellfish. Since that time, 39 closures of shellfish harvesting areas have occurred based on STX levels exceeding the international standard action level. However, natural *P. bahamense* populations are not routinely measured for STX production. *P. bahamense* populations in Puerto Rico have not been measured for toxicity and are presumed to be non-toxic (pers. comm., VCHT staff). Overall, this study consisted of a small sample size, and more widespread sampling would be needed to draw conclusions about the relative proportions of *sxtA4* genotypes in the areas sampled and the relationships of these results to the STX levels in the regions.

3.3. Gene Variants

Direct sequencing of *sxtA4* amplicons from single cells indicated multiple variants, evidenced by base substitutions that consistently occurred in an ca. 200 bp region (data not shown). Additionally, two distinct variants were consistently obtained from most cells, one of which displayed a 15 bp deletion.

Further testament as to *sxtA4* diversity within *P. bahamense* was evidenced by samples in which *sxtA4* was not initially detected. Single cells which yielded only an 18S rRNA gene amplicon in the initial multiplex were then screened with two additional PCRs: (1) the nested PCR using the 166F/680R primer set that targeted a conserved region of the *sxtA4* gene and (2) a PCR with only the sxtA4 F1/680R primers. A small subset of these samples yielded an amplicon indicative of *sxtA4* with the 166F/680R primers. Sequencing of these 500 bp amplicons confirmed *sxtA4*. Amplification with the conserved primer (166F) but not the *Pyrodinium*-specific (F1) suggests sequence variation in this region (F1 is located 165 bp upstream of 166F) of *P. bahamense*. In comparing *sxtA4* sequences available in GenBank that encompassed this region (all *Alexandrium* spp. and a single *Pyrodinium*) and gDNA and cDNA sequences of the toxic lab isolate, the F1 primer displayed sequence variation in *Alexandrium*, while the 166F primer was conserved across *Pyrodinium* and *Alexandrium* spp (see sequence alignments in Supplementary Materials).

To assess the extent of *sxtA4* variation within individual cells, the *sxtA4* amplicons from a small subset of single cells representative of the different sampling sites were cloned and sequenced. As with the direct sequencing of *sxtA4* PCR amplicons, two variants were consistently obtained from clones of all samples, with one of the variants displaying the 15 bp deletion (Figure 4). The *sxtA4* DNA sequences amplified and cloned from single cells were aligned with the cDNA sequences from the lab isolate. The 15 bp region was present in the cDNA copies of the lab isolate, indicating that this variant is likely a pseudogene. This was found to be the case with *A. fundyense* strains, in which a 63 bp deletion was found in some gDNA sequences but not the corresponding cDNA sequences [17].

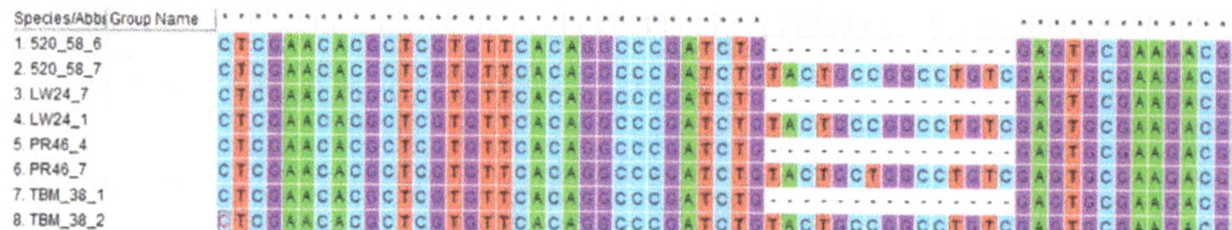

Figure 4. Two distinct variants were recovered from individual cells at all sampling sites, defined by a 15 bp deletion in one of the variants.

SxtA4 clone sequences from Puerto Rico showed two general variants, defined by the presence or absence of the 15 bp stretch. Overall, the percent identity among sequences ranged from 96–100%; minimal (2–5) base substitutions occurred among clones from the same general variant. *SxtA4* clones from the TBM site also showed the two variants. A greater diversity of sequences were obtained, with the percent identity spanning 91–98%. Both sequence variants were also obtained from site LW, and percent identity ranged from 95–99%. The two sxtA4 variants were also recovered from clones from 520, with percent identity ranging from 95–99%.

A neighbor-joining tree constructed with the cloned *sxtA4* amplicons from individual cells showed sequences formed two clusters; however, these clusters were not defined by sampling site/population but rather by the presence/absence of the 15 bp region (Figure 5). Comparison of the two representative sequences from the single cell clones with existing *Pyrodinium sxtA4* sequences showed the sequence variant not lacking the 15 bp segment was 99% identical to existing *P. bahamense sxtA4* sequences, and the variant with the 15 bp deletion was 95–96% identical (Table S2). This indicates the gene is conserved across populations, though this must be interpreted with caution due to the limited number of sequences and the length of the sequences.

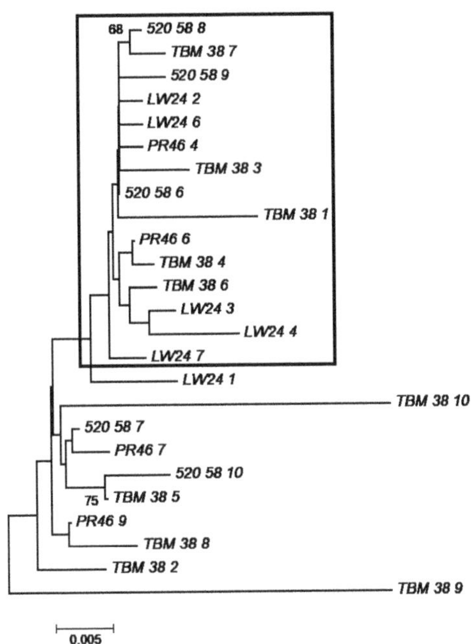

Figure 5. Neighbor Joining tree of *sxtA4* sequences obtained from clones of single cells of *P. bahamense* from the representative sampling sites. Percentages from the bootstrap test (500 replicates) are shown next to the branches. The Tamura 3-parameter method best fit the data and so was used to compute the evolutionary distances. Twenty-five nucleotide sequences were included in the analysis. All ambiguous positions were removed for each sequence pair. There were a total of 517 positions in the final dataset. Evolutionary analyses were conducted in MEGA7. Nomenclature is as follows: Sampling site_cell_clone number. Note that for each site, the "cell" number is the same, as all sequences from that site were derived from the same cell. DB = Diamond Bay (IRL); TBM = Telemar Bay Marina (IRL); LW = Lee Wenner Park (IRL); 520 = 520 Bridge (IRL); PR = Mosquito Bay in Puerto Rico. Sequences with the 15 bp deletion are boxed.

4. Conclusions

P. bahamense was previously classified as a single species with two varieties based on biochemical and morphological differences [56]. The Indo-Pacific variety was designated 'compressum', and the Atlantic-Caribbean variety 'bahamense'. However, both "varieties" have since been found in the same geographic locations, and one of the key distinguishing differences, the lack of toxin production by var. *bahamense*, has since been disproven. The varietal designation has been removed based on a recent reinvestigation of morphological attributes, and the suggestion that the presence of both varieties within the same plankton sample is likely the occurrence of different life stages [57]. Large subunit rRNA gene sequence analysis indicated the existence of Indo-Pacific and Atlantic-Caribbean ribotypes, leading to the suggestions that *P. bahamense* is a species complex. A previous study showed

18S rRNA gene sequences of *P. bahamense* from multiple locations in the IRL, Puerto Rico, and the Indo- Pacific to be nearly identical and forming a single cluster [48]. Comparison of 18S rRNA gene sequences from *sxtA4+* and *sxtA4-* genotype cells in this study were compared to existing 18S rRNA gene sequences from the Indo-Pacific and Atlantic-Caribbean. Not surprising, all sequences were 99.6–100% identical (SI).

A previous study by our group showed, at the single cell level, that two different luciferase (*lcf*, responsible for bioluminescence) variants were consistently recovered from *P. bahamense* populations in both Florida and Puerto Rico [48]. *Lcf* codes for a single polypeptide comprised of three homologous domains ("D1", "D2", "D3") [58,59]. The individual domains among species (i.e., D1 *A. tamarense*, D1 *A. affine*) are more similar than among the three different domains of the same species. The sequences amplified from individual cells from Florida and Puerto Rico formed two distinct clusters defined by a set of non-synonymous substitutions. Only one of the two types of *lcf* sequence, but not both, was recovered from each individual cell. Both types of *lcf* sequences were obtained from all sampling sites. Primers selectively amplified domain 3 (D3), indicating the two *P. bahamense lcf* sequences represented gene variants rather than amplification of the different domains. The two *P. bahamense* sequences are 87% similar. The variation between the two *P. bahamense lcf* sequences is much greater than that of bioluminescent species with known gene variants (i.e., *Pyrocystis lunula* possesses LcfA and LcfB, with 97% amino acid similarity). The significance underlying this variation remains unknown.

The results obtained here demonstrate that within small discrete natural populations (i.e., 3 L of surface water) of *P. bahamense* there exist both *sxtA4+* and *sxtA4-* genotypes. Correlation of *lcf* variant with *sxtA4* genotype may provide insights as to a potential *Pyrodinium* species complex. The next step will be determining how and if *sxtA4+/-* frequencies change over time, and the environmental factors driving these fluctuations. Understanding the factors driving bloom diversity and evolution is essential to understanding bloom dynamics and success over both space and time. This is the first study, to our knowledge, to examine the genetic potential for toxin production within natural dinoflagellate bloom sub-populations, which is critical for understanding bloom ecology and evolution.

Supplementary Materials: The following are available online at https://www.mdpi.com/article/10.3390/microorganisms9061128/s1, Detailed methods on additional screenings of *P. bahamense* cells following the initial multiplex PCR; sequence comparisons of *Pyrodinium* and *Alexandrium* illustrating sequence variability in sxtA4F1 and 166F primer binding sites; Table S1 and Methods for Detection of STX congeners in *P. bahamense* toxic lab isolate; Table S2: Percent Identity among *sxtA4 Pyrodinium bahamense* clones; representative 18S rRNA gene sequences used in *P. bahamense* analysis.

Author Contributions: Conceptualization, K.C.; methodology, K.C. and G.D.; formal analysis, K.C.; investigation, K.C.; resources, K.C.; data curation, K.C.; writing—original draft preparation, K.C.; writing—review and editing, K.C.; supervision, K.C.; project administration, K.C. All authors have read and agreed to the published version of the manuscript.

Funding: This research received no external funding and was funded by UMBC start-up funding to K.C.

Data Availability Statement: Sequences described in this study have been submitted to GenBank.

Acknowledgments: The authors thank Mark Martin, Vieques Conservation and Historical Trust, for collection of samples from Mosquito Bay and Tsetso Bachvaroff for insightful discussions. The authors thank Cary Lopez and staff at the Florida Fish & Wildlife Conservation Commission, Fish and Wildlife Research Institute for providing the P. bahamense isolate from the Indian River Lagoon.

Conflicts of Interest: The authors declare no conflict of interest.

References

1. Cusick, K.D.; Sayler, G.S. An overview on the marine neurotoxin saxitoxin: Genetics, molecular targets, methods of detection, and ecological functions. *Mar. Drugs* **2013**, *11*, 991–1018. [CrossRef]

2. Landsberg, J.H.; Hall, S.; Johannessen, J.N.; White, K.D.; Conrad, S.M.; Abbott, J.P.; Flewelling, L.J.; Richardson, R.W.; Dickey, R.W.; Jester, E.L.E.; et al. Saxitoxin puffer fish poisoning in the United States, with the first report of Pyrodinium bahamense as the putative toxin source. *Environ. Health Persp.* **2006**, *114*, 1502–1507. [CrossRef]
3. Anderson, D.M.; Kulis, D.M.; Sullivan, J.J.; Hall, S.; Lee, C. Dynamics and physiology of saxitoxin production by the dinoflagellates Alexandrium spp. *Mar. Biol.* **1990**, *104*, 511–524. [CrossRef]
4. Hallegraeff, G.M.; Steffensen, D.A.; Wetherbee, R. Three estuarine Australian dinoflagellates that can produce paralytic shellfish toxins. *J. Plankton Res.* **1988**, *10*, 533–541. [CrossRef]
5. Harada, T.; Oshima, Y.; Kamiya, H.; Yasumoto, T. Confirmation of paralytic shellfish toxins in the dinoflagellate pyrodinium-bahamense var. compressa and bivalves in Palau. *Bull. Jpn. Soc. Sci. Fish.* **1982**, *48*, 821–825. [CrossRef]
6. Oshima, Y.; Hasegawa, M.; Yasumoto, T.; Hallegraeff, G.; Blackburn, S. Dinoflagellate Gymnodinium catenatum as the source of paralytic shellfish toxins in tasmanian shellfish. *Toxicon* **1987**, *25*, 1105–1111. [CrossRef]
7. Carmichael, W.W.; Evans, W.R.; Yin, Q.Q.; Bell, P.; Moczydlowski, E. Evidence for paralytic shellfish poisons in the freshwater cyanobacterium Lyngbya wollei (Farlow ex Gomont) comb. *Nov. Appl. Environ. Microbiol.* **1997**, *63*, 3104–3110. [CrossRef]
8. Negri, A.P.; Jones, G.J. Bioaccumulation of paralytic shellfish poisoning (PSP) toxins from the cyanobacterium Anabaena circinalis by the freshwater mussel Alathyria condola. *Toxicon* **1995**, *33*, 667–678. [CrossRef]
9. Shimizu, Y. Microalgal metabolites. *Chem. Rev.* **1993**, *93*, 1685–1698. [CrossRef]
10. Shimizu, Y. Microalgal metabolites: A new perspective. *Annu. Rev. Microbiol.* **1996**, *50*, 431–465. [CrossRef] [PubMed]
11. Kellmann, R.; Mihali, T.K.; Jeon, Y.J.; Pickford, R.; Pomati, F.; Neilan, B.A. Biosynthetic intermediate analysis and functional homology reveal a saxitoxin gene cluster in cyanobacteria. *Appl. Environ. Microbiol.* **2008**, *74*, 4044–4053. [CrossRef]
12. Mihali, T.K.; Kellmann, R.; Neilan, B.A. Characterisation of the paralytic shellfish toxin biosynthesis gene clusters in Anabaena circinalis AWQC131C and Aphanizomenon sp. NH-5. *BMC Biochem.* **2009**, *10*, 8. [CrossRef]
13. Mihali, T.K.; Carmichael, W.W.; Neilan, B.A. A putative gene cluster from a Lyngbya wollei bloom that encodes paralytic shellfish toxin biosynthesis. *PLoS ONE* **2011**, *6*, e14657. [CrossRef]
14. Stuken, A.; Orr, R.J.S.; Kellmann, R.; Murray, S.A.; Neilan, B.A.; Jakobsen, K.S. Discovery of nuclear-encoded genes for the neurotoxin saxitoxin in dinoflagellates. *PLoS ONE* **2011**, *6*, e20096. [CrossRef]
15. Hackett, J.D.; Wisecarver, J.H.; Brosnahan, M.L.; Kulis, D.M.; Anderson, D.M.; Bhattacharya, D.; Plumley, F.G.; Erdner, D.L. Evolution of saxitoxin synthesis in cyanobacteria and dinoflagellates. *Mol. Biol. Evol.* **2013**, *30*, 70–78. [CrossRef]
16. Kellmann, R.; Michali, T.K.; Neilan, B.A. Identification of a saxitoxin biosynthesis gene with a history of frequent horizontal gene transfers. *J. Mol. Evol.* **2008**, *67*, 526–538. [CrossRef]
17. Murray, S.A.; Diwan, R.; Orr, R.J.S.; Kohli, G.S.; John, U. Gene duplication, loss, and selection in the evolution of saxitoxin biosynthesis in alveolates. *Mol. Phylogenetics Evol.* **2015**, *92*, 165–180. [CrossRef]
18. Wang, H.; Kim, H.; Ki, J.-S. Transcriptome survey and toxin measurements reveal evolutionary modification and loss of saxitoxin biosynthesis genes in the dinoflagellates Amphidinium carterae and Prorocentrum micans. *Ecotoxicol. Environ. Saf.* **2020**, *195*, 110474. [CrossRef]
19. Murray, S.A.; Wiese, M.; Stuken, A.; Brett, S.; Kellmann, R.; Hallegraeff, G.; Neilan, B.A. SxtA-based quantitative molecular assay to identify saxitoxin-producing harmful algal blooms in marine waters. *Appl. Environ. Microbiol.* **2011**, *77*, 7050–7057. [CrossRef]
20. Orr, R.J.S.; Stuken, A.; Murray, S.A.; Jakobsen, K.S. Evolutionary acquisition and loss of saxitoxin biosynthesis in dinoflagellates: The second "core" gene, sxtG. *Appl. Environ. Microbiol.* **2013**, *79*, 2128–2136. [CrossRef]
21. Murray, S.A.; Hoppenrath, M.; Orr, R.J.S.; Bolch, C.; John, U.; Diwan, R.; Yauwenas, R.; Harwood, T.; de Salas, M.; Neilan, B.; et al. Alexandrium diversaporum sp. nov., a new non-saxitoxin producing species: Phylogeny, morphology and sxtA genes. *Harmful Algae* **2014**, *31*, 54–65. [CrossRef]
22. Touzet, N.; Franco, J.M.; Raine, R. Characterization of nontoxic and toxin-producing strains of Alexandrium minutum (Dinophyceae) in Irish coastal waters. *Appl. Environ. Microbiol.* **2007**, *73*, 3333–3342. [CrossRef]
23. Alpermann, T.J.; Tillmann, U.; Beszteri, B.; Cembella, A.D.; John, U. Phenotypic variation and genotypic diversity in a planktonic population of the toxigenic marine dinoflagellate Alexandrium tamarense (Dinophyceae). *J. Phycol.* **2010**, *46*, 18–32. [CrossRef]
24. Menden-Deuer, S.; Montalbano, A.L. Bloom formation potential in the harmful dinoflagellate Akashiwo sanguinea: Clues from movement behaviors and growth characteristics. *Harmful Algae* **2015**, *47*, 75–85. [CrossRef]
25. Anderson, D.M.; Alpermann, T.J.; Cembella, A.D.; Collos, Y. Masseret E, Montresor M. The globally distributed genus Alexandrium: Multifaceted roles in marine ecosystems and impacts on human health. *Harmful Algae* **2012**, *14*, 10–35. [CrossRef]
26. Whittaker, K.A.; Rignanese, D.R.; Olson, R.J.; Rynearson, T.A. Molecular subdivision of the marine diatom Thalassiosira rotulain relation to geographic distribution, genome size, and physiology. *BMC Evol. Biol.* **2012**, *12*, 209. [CrossRef]
27. Saravanan, V.; Godhe, A. Genetic heterogeneity and physiological variation among seasonally separated clones of Skeletonema marinoi (Bacillariophyceae) in the Gullmar Fjord, Sweden. *Eur. J. Phycol.* **2010**, *45*, 177–190. [CrossRef]
28. Rynearson, T.A.; Armbrust, E.V. Genetic differentiation among populations of the planktonic marine diatim Ditylum brightwellii (Bacillariophyceae). *J. Phycol.* **2004**, *40*, 34–43. [CrossRef]
29. Richlen, M.L.; Erdner, D.L.; McCauley, L.A.R.; Libera, K.; Anderson, D.M. Extensive genetic diversity and rapid population differentiation during blooms of Alexandrium fundyense (Dinophyceae) in an isolated salt pond on Cape Cod, MA, USA. *Ecol. Evol.* **2012**, *2*, 2588–2599. [CrossRef]

30. Hallegraeff, G.M.; Blackburn, S.I.; Doblin, M.A.; Bolch, C.J.S. Global toxicology, ecophysiology and population relationships of the chainforming PST dinoflagellate Gymnodinium catenatum. *Harmful Algae* **2012**, *14*, 130–143. [CrossRef]

31. Morquecho, L. Pyrodinium bahamense, one the most significant harmful dinoflagellates in Mexico. *Front. Mar. Sci.* **2019**, *6*, 1–8. [CrossRef]

32. Phlips, E.J.; Badylak, S.; Christman, M.; Wolny, J.; Brame, J.; Garland, J.; Hall, L.; Hart, J.; Landsberg, J.; Lasi, M.; et al. Scales of temporal and spatial variability in the distribution of harmful algae species in the Indian River Lagoon, Florida, USA. *Harmful Algae* **2011**, *10*, 277–290. [CrossRef]

33. Usup, G.; Ahmada, A.; Matsuoka, K.; Lim, P.T.; Leaw, C.P. Biology, ecology and bloom dynamics of the toxic marine dinoflagellate Pyrodinium bahamense. *Harmful Algae* **2012**, *14*, 301–312. [CrossRef]

34. Montojo, U.M.; Sakamoto, S.; Cayme, M.F.; Gatdula, N.C.; Furio, E.F.; Relox, J.J.R.; Shigeru, S.; Fukuyo, Y.; Kodama, M. Remarkable difference in accumulation of paralytic shellfish poisoning toxins among bivalve species exposed to Pyrodinium bahamense var. compressum bloom in Masinloc bay, Philippines. *Toxicon* **2006**, *48*, 85–92. [CrossRef] [PubMed]

35. Llewellyn, L.; Negri, A.; Robertson, A. Paralytic shellfish toxins in tropical oceans. *Toxin Rev.* **2006**, *25*, 159–196. [CrossRef]

36. Azanza, R.V.; Miranda, L.N. Phytoplankton composition and Pyrodinium bahamense toxic blooms in Manila Bay, Philippines. *J. Shellfish Res.* **2001**, *20*, 1251–1255.

37. Azanza, R.V.; Taylor, F.J.R. Are Pyrodinium blooms in the Southeast Asian region recurring and spreading? A view at the end of the millennium. *AMBIO* **2001**, *30*, 356–364. [CrossRef]

38. Gacutan, R.Q.; Tabbu, M.Y.; Aujero, E.J.; Icatlo, F. Paralytic shellfish poisoning due to pyrodinium-bahamense var compressa in Mati, Davao-Oriental, Philippines. *Mar. Biol.* **1985**, *87*, 223–227. [CrossRef]

39. Bodager, D. Outbreak of saxitoxin illness following consumption of Florida pufferfish. *Fl J. Environ. Health* **2002**, *179*, 9–13.

40. Brandenburg, K.M.; Wohlrab, S.; John, U.; Kremp, A.; Jerney, J.; Krock, B.; Van de Waal, D.B. Intraspecific trait variation and trade-offs within and across populations of a toxic dinoflagellate. *Ecol. Lett.* **2018**, *21*, 1561–1571. [CrossRef]

41. Murray, S.A.; Mihali, T.K.; Neilan, B.A. Extraordinary conservation, gene loss, and positive selection in the evolution of an ancient neurotoxin. *Mol. Biol. Evol.* **2011**, *28*, 1173–1182. [CrossRef]

42. Zhang, Y.; Zhang, S.F.; Lin, L.; Wang, D.Z. Comparative transcriptome analysis of a toxin-producing dinoflagellate Alexandrium catenella and its non-toxic mutant. *Mar. Drugs* **2014**, *12*, 5698–5718. [CrossRef]

43. John, U.; Beszteri, B.; Derelle, E.; Van de Peer, Y.; Read, B.; Moreau, H.; Cembella, A.D. Novel insights into evolution of protistan polyketide synthases through phylogenomic analysis. *Protist* **2008**, *158*, 21–30. [CrossRef]

44. Eichholz, K.; Beszteri, B.; John, U. Putative monofunctional type I polyketide synthase units: A dinoflagellate-specific feature? *PLoS ONE* **2012**, *7*, e48624. [CrossRef]

45. Nosenko, T.; Bhattacharya, D. Horizontal gene transfer in chromalveolates. *BMC Evol. Biol.* **2007**, *7*, 173. [CrossRef]

46. Wisecaver, J.H.; Brosnahan, M.L.; Hackett, J.D. Horizontal gene transfer is a significant driver of gene innovation in dinoflagellates. *Genome Biol. Evol.* **2013**, *5*, 2368–2381. [CrossRef]

47. Altschul, S.F.; Gish, W.; Miller, W.; Myers, E.W.; Lipman, D.J. Basic local aligment search tool. *J. Mol. Biol.* **1990**, *215*, 403–410. [CrossRef]

48. Cusick, K.D.; Wilhelm, S.W.; Hargraves, P.E.; Sayler, G.S. Single-cell PCR of the luciferase conserved catalytic domain reveals a unique cluster in the toxic bioluminescent dinoflagellate Pyrodinium bahamense. *Aquat. Biol.* **2016**, *25*, 139–150. [CrossRef]

49. Tomas, C.R. (Ed.) *Identifying Marine Phytoplankton*; Academic Press: San Diego, CA, USA, 1997; p. 858.

50. Zhang, Z.; Schwartz, S.; Wagner, L.; Miller, W. A greedy algorithm for aligning DNA sequences. *J. Comput. Biol.* **2000**, *7*, 203–214. [CrossRef]

51. Phlips, E.J.; Badylak, S.; Grosskopf, T. Factors affecting the abundance of phytoplankton in a restricted subtropical lagoon, the Indian River Lagoon, Florida, USA. *Estuar. Coast. Shelf Sci.* **2002**, *55*, 385–402. [CrossRef]

52. Badylak, S.; Phlips, E.J. Spatial and temporal patterns of phytoplankton composition in a subtropical coastal lagoon, the Indian River Lagoon, Florida, USA. *J. Plankton Res.* **2004**, *26*, 1229–1247. [CrossRef]

53. Phlips, E.J.; Badylak, S.; Christman, M.C.; Lasi, M.A. Climatic trends and temporal patterns of phytoplankton composition, abundance and succession in the Indian River Lagoon, Florida, USA. *Estuar. Coast.* **2010**, *33*, 498–512. [CrossRef]

54. Phlips, E.J.; Badylak, S.; Bledsoe, E.; Cichra, M. Factors affecting the distribution of Pyrodinium bahamense var. bahamense in coastal waters of Florida. *Mar. Ecol. Prog. Ser.* **2006**, *322*, 99–115. [CrossRef]

55. Grasso, S.; Albrecht, M.; Bras, M.M. Seasonal abundance of Pyrodinium bahamense (order Peridiniales, family Gonyaulacaceae) in Mosquito Bay, Vieques, Puerto Rico. *J. Coast Life Med.* **2016**, *4*, 277–283. [CrossRef]

56. Steidinger, K.A.; Tester, L.S.; Taylor, F.J.R. A redescription of Pyrodinium bahamense var. compressa (Bohm) stat. nov. from Pacific red tides. *Phycologia* **1980**, *19*, 329–337. [CrossRef]

57. Mertens, K.N.; Wolny, J.; Carbonell-Moore, C.; Bogus, K.; Ellegaard, M.; Limoges, A.; de Vernal, A.; Gurdebeke, P.; Omura, T.; Al-Muftah, A.; et al. Taxonomic re-examination of the toxic armored dinoflagellate Pyrodinium bahamense Plate 1906: Can morphology or LSU sequencing separate P. bahamense var. compressum from var. bahamense? *Harmful Algae* **2015**, *41*, 1–24. [CrossRef]

58. Liu, L.; Wilson, T.; Hastings, J.W. Molecular evolution of dinoflagellate luciferases, enzymes with three catalytic domains in a single polypeptide. *Proc. Natl. Acad. Sci. USA* **2004**, *101*, 16555–16560. [CrossRef]

59. Okamoto, O.K.; Liu, L.; Robertson, D.L.; Hastings, J.W. Members of a dinoflagellate luciferase gene family differ in synonymous substitution rates. *Biochemistry* **2001**, *40*, 15862–15868. [CrossRef]

 *microorganisms*

MDPI

Article

Screening a Spliced Leader-Based *Symbiodinium microadriaticum* cDNA Library Using the Yeast-Two Hybrid System Reveals a Hemerythrin-Like Protein as a Putative SmicRACK1 Ligand

Tania Islas-Flores [1,*], Edgardo Galán-Vásquez [2] and Marco A. Villanueva [1,*]

[1] Unidad Académica de Sistemas Arrecifales, Instituto de Ciencias del Mar y Limnología, Universidad Nacional Autónoma de México, UNAM, Prolongación Avenida Niños Héroes S/N, Puerto Morelos, Quintana Roo 77580, México

[2] Departamento de Ingeniería de Sistemas Computacionales y Automatización, Instituto de Investigación en Matemáticas Aplicadas y en Sistemas, Universidad Nacional Autónoma de México, UNAM, Circuito Escolar 3000, Ciudad Universitaria, Ciudad de México CP 04510, México; edgardo.galan@iimas.unam.mx

* Correspondence: tislasf@gmail.com (T.I.-F.); marco@cmarl.unam.mx (M.A.V.); Tel.: +52-998-871-0009 (T.I.-F. & M.A.V.)

Citation: Islas-Flores, T.; Galán-Vásquez, E.; Villanueva, M.A. Screening a Spliced Leader-Based *Symbiodinium microadriaticum* cDNA Library Using the Yeast-Two Hybrid System Reveals a Hemerythrin-Like Protein as a Putative SmicRACK1 Ligand. *Microorganisms* **2021**, *9*, 791. https://doi.org/10.3390/microorganisms9040791

Academic Editor: Shauna Murray

Received: 25 February 2021
Accepted: 16 March 2021
Published: 9 April 2021

Publisher's Note: MDPI stays neutral with regard to jurisdictional claims in published maps and institutional affiliations.

Abstract: The dinoflagellate Symbiodiniaceae family plays a central role in the health of the coral reef ecosystem via the symbiosis that establishes with its inhabiting cnidarians and supports the host metabolism. In the last few decades, coral reefs have been threatened by pollution and rising temperatures which have led to coral loss. These events have raised interest in studying Symbiodiniaceae and their hosts; however, progress in understanding their metabolism, signal transduction pathways, and physiology in general, has been slow because dinoflagellates present peculiar characteristics. We took advantage of one of these peculiarities; namely, the post-transcriptional addition of a Dino Spliced Leader (Dino-SL) to the 5′ end of the nuclear mRNAs, and used it to generate cDNA libraries from *Symbiodinium microadriaticum*. We compared sequences from two Yeast-Two Hybrid System cDNA Libraries, one based on the Dino-SL sequence, and the other based on the SMART technology (Switching Mechanism at 5′ end of RNA Transcript) which exploits the template switching function of the reverse transcriptase. Upon comparison of the performance of both libraries, we obtained a significantly higher yield, number and length of sequences, number of transcripts, and better 5′ representation from the Dino-SL based library than from the SMART library. In addition, we confirmed that the cDNAs from the Dino-SL library were adequately expressed in the yeast cells used for the Yeast-Two Hybrid System which resulted in successful screening for putative SmicRACK1 ligands, which yielded a putative hemerythrin-like protein.

Keywords: cDNA library; coral reefs; Hemerythrin-like protein; RACK1; spliced leader; *Symbiodinium*; yeast two-hybrid

1. Introduction

Symbiodiniaceae are highly diverse dinoflagellate algae that establish symbiosis with cnidarians and marine organisms of diverse phyla including corals. Coral reefs shelter a highly biodiverse ecosystem, which depends on the functional symbiosis with dinoflagellate members of such a family [1,2]. The dinoflagellates reside in symbiosomal membranes in the host gastrodermal cells, where they carry out the photosynthetic process that fuels the productivity and diversity of the coral reef ecosystems. In the past decades, coral reefs have suffered large areal losses due to coral death caused by diseases and coral bleaching, the latter being a cause of the symbiosis breakdown; however, the mechanisms underlying the metabolism, symbiont selection and establishment, maintenance, and disruption of the symbiosis, are still poorly understood. The slow progress in the study of cnidarian–dinoflagellate symbiosis is a consequence of the scarcely available molecular and functional genomic tools applied to corals and Symbiodiniaceae. To date, it has not been possible to obtain silenced, mutated, or knockout stable lines of Symbiodiniaceae to carry out integral functional genomic studies that may shed light on these pathways [3,4].

Symbiodiniaceae present genomic features that make them particularly interesting: (a) they possess some of the largest nuclear genomes among eukaryotes; and (b) their DNA is arranged on permanently condensed liquid-crystalline chromosomes. This way of organization has been proposed as a third state of chromosomal folding besides the nucleosomal and the supercoiled circular arrangements in eukaryotes and bacteria [5,6]. A third and significant genomic feature is that their transcripts contain a 22-nt conserved sequence called Dino Spliced Leader (Dino-SL), DCCGUAGCCAUUUUGGCUCAAG (D = U, A or G), which is post-transcriptionally added to the 5′ end of all their nuclear mRNAs [7]. The spliced leader gene has an intronic region with unknown function, and an exonic region which is transferred (*trans*-spliced) to the 5′ pre-mRNAs [8]. The *trans*-splicing of the SL sequence is part of an mRNA processing mechanism, in which a small fragment of a non-coding RNA is transferred to a splice acceptor at the 5′-UTR of pre mRNAs [9]. The Dino-SL sequence has proved to be useful as a molecular tag that allows us to: a) specifically select dinoflagellate mRNA from mixed microbial samples [10]; (b) construct cDNA libraries from environmental samples using a Dino-SL based approach with high specificity [10,11]; and (c) identify any dinoflagellate gene [12]. These approaches have been useful to obtain specific cDNA libraries from dinoflagellates; however, it has not been explored whether they can be used for accomplishing high-quality cDNA libraries that would allow to collect full-length cDNA clones efficiently [13]. The actual methodology to construct a cDNA library involves the synthesis of second-strand cDNA, which can produce artifacts, including the loss of information at the 5′ and 3′ ends of mRNAs where gene regulation information is encoded [14]; however, this step is necessary to avoid the loss of less-abundant transcripts.

The Dino-SL is a convenient tag that allows to prime the 5′ end of the dinoflagellate mRNA, in addition to the dT at the 3′ end. Therefore, it is expected that, when using the Dino-SL tag, cDNA libraries representing a higher number of complete transcripts as well as a higher representation of less-abundant transcripts, will be obtained. The result would be a molecular tool allowing to analyze the sequencing data obtained from them, with more reliability than from conventionally constructed libraries. Furthermore, it opens the opportunity to generate cDNA libraries for applications like the Yeast-Two Hybrid System (Y2HS) that can be used to identify ligands of proteins of interest.

The Y2HS is one of the most widely used methods for in vivo identification of interacting proteins, confirming interactions between two known proteins, and/or mapping interacting domains; in addition, it has been used to successfully map interaction networks on a large scale [15]. In this technique, the interaction between two proteins of interest is detected via the reconstitution of a transcription factor and the subsequent activation of reporter genes under the control of a transcription factor [16]. It exploits the fact that the DNA-binding domain (BD) of GAL4 is incapable of activating transcription unless physically associated with an activating domain (AD) [17]. In a two-hybrid assay, a protein of interest is fused to a DNA binding domain (BD), then transfected to a yeast strain; this construct is called the "bait". In order to be useful, the "bait" must not be able to activate the transcription of the reporter(s) on its own, so that a cDNA library that is fused to the activation domain (AD) called the "prey" can be screened. Thus, when "prey" library clones express proteins capable of interacting with the "bait", they can be identified by their ability to activate the reporter(s). Although this system presents an opportunity to identify or confirm protein–protein interactions, it is not a failsafe approach [17]. Success in the identification of ligands by the Y2HS depends on several factors that can negatively affect the detection of protein–protein interactions. For example, the use of chimeras is a concern because it can change the protein ("bait" and/or "prey") conformations and thus disrupt their interaction sites; another drawback is the use of yeast as a host to express the cDNA library of heterologous systems, which implies that some required post-translational modifications for protein–protein interactions could be incorrectly carried out or not occurring at all. Finally, cDNA libraries with underestimated or incomplete transcripts could lead, in turn, to incomplete, untranslated (incorrect frame), or low amount of "prey" proteins. The Dino-SL based cDNA library approach could overcome all those problems. In this work, in an attempt to obtain a library with higher complexity than the conventionally generated SMART (Switching Mechanism at 5′ end of RNA Transcript) libraries, we sought to generate a cDNA library of *Symbiodinium microadriaticum* ssp. *microadriaticum* using the Dino-SL strategy (DINO library) to amplify the dinoflagellate cDNA and clone it into the pGADT7-rec plasmid. Then, to confirm that with this approach we could indeed obtain a longer and higher number of transcripts, we compared it to a conventionally generated cDNA library (SMART library). The DINO library allowed us to obtain 54% more transcripts with a $\geq$5000 bp length, than with the SMART library. In order to validate the DINO library as a tool for the Y2HS, we used the Receptor for Activated C Kinase 1 from *S. microadriaticum* (SmicRACK1) as bait to screen the library for prey transcripts and obtained a sequence that upon BLAST [18] matched a hemerythrin-like protein as a putative SmicRACK1 ligand. This was further confirmed by direct activation of the reporter gene by both sequences in a second Y2H assay. Thus, the Dino-SL strategy provides a molecular tool to obtain high throughput cDNA libraries that can be used as expression libraries for molecular approaches.

2. Materials and Methods

2.1. Cell Lines and Cultures

Aseptic cultures of the photosynthetic dinoflagellate *Symbiodinium microadriaticum* ssp. *microadriaticum* (MAC-CassKB8), originally isolated from the jellyfish *Cassiopea xamachana*, and kindly donated by Dr. Mary Alice Coffroth (State University of New York at Buffalo), were routinely maintained as in vitro culture in ASP-8A medium [19] under standard photoperiod cycles of 12 h light/dark with a light regime of 80 μmoles quanta/m^2 s, at 25 °C. The yeast strain Y187 was used as host for both "prey" cDNA libraries. This strain is stored in our laboratory as glycerol stocks at -80 °C, and also maintained on PDA medium at 4 °C.

2.2. Total RNA Isolation, cDNA Synthesis and cDNA Amplification

Total RNA from exponentially growing *S. microadriaticum* cultures (1×10^5 cells/mL) was extracted as follows. Cultures were harvested by centrifugation at $1400\times g$ for 5 min, then the pellet was suspended with TRI REAGENT (Sigma, St. Louis, MO, USA) and vigorously shaken at 48 rpm with glass beads (425–600 μm; Sigma) in a bead beater (Biospec, Bartlesville, OK, USA) for 2–3 min at 4 °C; total RNA was extracted according to the manufacturer's instructions. The RNA was further purified with the RNA CLEANUP Kit (Qiagen, Valencia, CA, USA), and mRNA was purified with the OLIGOTEX kit (Qiagen, Valencia, CA, USA) according to the protocol in the accompanying user manual. All the RNA samples were treated with RNase-free DNase I (Thermo Fisher Scientific, Waltham, MA, USA) to eliminate DNA. The integrity of total RNA was analyzed by 1.2% denaturing agarose gel electrophoresis, followed by GELRED staining (Biotium, Fremont, CA, USA). The concentration and purity of the total RNA were determined with a SmartSpec 3000 spectrophotometer (BioRad, Hercules, CA, USA) at 260 and 280 nm, respectively. The cDNA synthesis was made with the Matchmaker Library Construction kit (Clontech Laboratories Inc., Mountain View, CA, USA), according to the manufacturer's instructions. Two tubes with 0.5–1 μg of mRNA were mixed with 1 μL 10 μM CDSIII oligo (5′ ATTCTAGAGGCCGAGGCGGCCGACATG-d(T)30VN 3′) and 2 μL of nuclease-free water. The tubes were incubated at 72 °C for 5 min and then quickly cooled on ice before their content was added to two tubes containing 2 μL of 5X first strand buffer, 1 μL 0.1 M DTT, 1 μL 10 mM dNTP's mix, and 1 μL of the SMART Moloney murine leukemia virus reverse transcriptase (M-MLV RT). Then, the mixtures were incubated as follows: 42 °C for

10 min, followed by 1 μL of SMARTIII oligo (5′ AAGCAGTGGTATCAACGCAGAGTGGC-CATTATGGCCGGG 3′) to only one tube; the incubation continued for 1 h at 42 °C; 75 °C for 10 min; and 23 °C for 5 min. At this point, 1 μL of RNAse H was added to both tubes; finally, the tubes were incubated at 37 °C for 20 min.

Long distance-polymerase chain reaction (LD-PCR) was carried out with the ADVAN-TAGE 2 POLYMERASE MIX (Clontech) following the manufacturer's instructions. For each cDNA (synthesized with or without the SMARTIII oligo), two 100 μL reactions were carried out. The reactions consisted of a mixture of 10 μL 10X Advantage 2 PCR Buffer, 2 μL 50X dNTPs mix, 2 μL 10 μM 3′ PCR oligo (5′ GTATCGATGCCCACCCTCTAGAGGC-CGAGGCGGCCGACA 3′), 10 μL 10X fusion solution, 2 μL 50X Advantage 2 polymerase mix, and 70 μL nuclease-free water along with: (a) 2 μL first strand cDNA synthesized with SMARTIII oligo and 2 μL 10 μM 5′ PCR oligo (5′ TTCCACCCAAGCAGTGGTAT-CAACGCAGAGTGG 3′); or (b) 2 μL first strand cDNA synthesized without SMARTIII oligo and 2 μL 10 μM DINO-SL oligo (5′ TTCCACCCAAGCAGTGGTATCAACGCA-GAGTGGCCATTATGGCCCCGTAGCCATTTTGGCTCAAG 3′). Then, the reactions were subjected to the following program: step 1, 94 °C, 30 s; step 2, 94 °C, 10 s; step 3, 68 °C, 6 min; 25 cycles from step 2–3. To confirm that the amplification took place (which should be seen as a smear on the gel), 7 μL of each reaction were analyzed on a 1.2% agarose gel subsequently stained with GelRed (Biotium, Fremont, CA, USA). The PCR products were purified with CHROMA SPIN^{TM} +TE-400 columns to discard DNA molecules of <200 pb.

2.3. Generation of cDNA Libraries

Two to five μg from each ds cDNA pool were mixed with 3 μg of *Sma*I linearized pGADT7-rec vector, and Y187 competent cells were transformed at library scale using the Yeastmaker Yeast Transformation System 2 (Clontech) according to the protocol in the accompanying user manual. The SMART and Dino-SL (DINO) libraries had 7.2×10^8 and 2.43×10^9 independent clones, respectively, which was an indication of the library complexity (which should be no less than 1×10^6 independent clones).

2.4. Sequencing of Libraries

To determine that the Dino-SL based approach improves the cDNA library with respect to the use of the regular methodology, we decided to sequence both Y2HS (DINO and SMART) libraries. A 1 mL aliquot from each library was thawed in a water bath at 25 °C. The plasmids from the libraries were purified with the ZYMOPREP YEAST PLASMID MINIPREP II kit (Zymo Research, Irvine, CA, USA). The libraries were amplified by LD-PCR with the ADVANTAGE 2 POLYMERASE MIX (Clontech) using the AD-Screen Fwd oligo (5′ CTATTCGATGATGAAGATACCCCACCAAACCCA 3′), and the AD-Screen Rv oligo (5′ GTGAACTTGCGGGGTTTTTCAGTATCTACGATT 3′) under the same conditions of cDNA amplification. The PCR products were sent for sequencing to the Mass Sequencing and Bioinformatics facility of the Institute of Biotechnology at UNAM. Sequencing was carried out by Illumina (2 × 75 cycles, 10 million reads).

2.5. Bioinformatic Analysis

Read quality control and cleaning were carried out using Fastp (v0.19.5), which can perform quality control, adapter trimming, quality filtering, per-read quality pruning and many other operations with a single scan of the FASTQ data. Default parameters were used [20]. The de novo assembly was carried out using Trinity (v2.8.4) [21], from the sequences free of: adapters, overrepresented sequences, low quality and low complexity. Then, TransDecoder was used to identify candidate coding regions from the assembled transcripts (v5.5.0). The candidate coding regions were used for BLAST analysis [18] against the draft of *S. microadriaticum* genome [22,23], and homologs were accepted if they had an e-value $< 1 \times 10^{-5}$, sequence identity > 90%, and alignment length > 90% of the individual proteins. The presence of conserved domains in the assembled transcripts was identified and annotated using HMMER [24]. KEGG pathway analysis of the candidate coding regions were performed using GhostKOALA [25].

Finally, the Kozak sequence for *Symbiodinium* (RCCATGGCN) was mapped using the DNA-pattern program which allows to search all occurrences of a pattern within DNA sequence [26]. The searches were done only in the direct strands, machining positions can be calculated either relative to the sequence start and 1 substitution were allowed.

2.6. Yeast-Two Hybrid System (Y2HS) Screening

In order to probe whether the DINO library was suitable to screen for protein interactions, we made a screen using the AH109 yeast strain harboring the bait construction pGBKT7-*SmicRACK1* as bait, and the DINO library as the prey in the Y187 yeast strain. The screening was carried out with the MATCHMAKER GAL4 Two-Hybrid System 3 kit following the instructions of the manufacturer (Clontech).

3. Results

3.1. The DINO Library Has Increased Yields and Higher Representation

The yield of the library is an important indicator of the transcript representation. Because the cDNA content quality could affect the performance of the libraries, we determined and compared the yield of each library to confirm differences in their complexity. Two *S. microadriaticum* full-length cDNA libraries were generated, one based on the SMAR-TIII oligo and the other one based on the Dino-SL tag. Both libraries had similar yield; the DINO library had 1.8×10^8 ufc/mL while the SMART library 1.6×10^8 ufc/mL. However, the number of independent clones was higher in the DINO (2.43×10^9) than in the SMART library (7.2×10^8) (Table 1). This indicates a favorable representation of transcripts in both libraries albeit with a better representation in the Dino-SL based library.

Table 1. Obtained yields from the DINO (Dino Spliced Leader) and SMART (Switching Mechanism at 5′ end of RNA Transcript) libraries.

	Recommended Minimum	DINO	SMART
Density cells/mL	$>2 \times 10^7$	6×10^8	5.9×10^8
UFC/mL	$>1 \times 10^7$	1.8×10^8	1.6×10^8
Independent clones n:	$>1 \times 10^6$	2.43×10^9	7.2×10^8

3.2. The Dino Spliced Leader (DINO) Library Presents Higher Transcript Identification

Because the Dino-SL sequence and a dT-based oligo were used as forward and reverse primers, respectively, to amplify the cDNA for the DINO library, it was expected that it was represented by a higher number of complete transcripts than the SMART library. The highest number of reads filtered with Fastp (v0.19.5) was obtained from the DINO library (35,311,534) compared with the number of reads obtained from the SMART library (17,715,026; Table 2); the sequenced reads had a mean length of 70 and 69 bp for the DINO and SMART libraries, respectively. After the de novo assembly, the sequences >1000 bp for the DINO library were 1.39-fold higher in the DINO library (1,168,986) than those in the SMART library (840,010) (Figure 1). Similarly, the sequences harboring >5000 bp were ~2.2-fold higher in the DINO library (89,797) than those in the SMART library (40,817) (Figure 1). The calculated N50 scaffold for the DINO library was 5671, while that for the SMART library was 3611 (Table 2). The de novo assembly produced 4760 transcripts from the DINO library, and 3611 transcripts from the SMART library, respectively (Table 2). The transcripts were used for the detection of coding sequences (CDS) and prediction of proteins resulting in the identification of 3117 CDS and 3117 proteins in the DINO library, and 2025 CDS and 2025 proteins in the SMART library (Table 2). These data indicated that the cDNA libraries based on Dino-SL allowed us to obtain a significant information gain.

Table 2. Read and assembly statistics for DINO and SMART libraries. Overview of the sequencing data, assembly, clustering, and annotation.

	DINO Library	**SMART Library**
Total reads (FastP v0.19.5)	35.311534 M [1]	17.715026 M
Total bases	2.479454 Gb [2]	1.236204 Gb
Transcripts (Trinity)	4760	3611
Total length contigs after TransDecoder		
N50 [3]	5671	4840
N75 [4]	4270	3534
GC content (%)	40.93	35.78
CDS [5]	3117	2025
Proteins	3117	2025
Orthologs	1422	1070

[1] M: millions; [2] Gb: gigabases; [3] N50: sequence length of the shortest contig at 50% of total genome length; [4] N75: same as N50 but length of shortest contig at 75%; [5] CDS: coding sequence.

Total sequences from the de novo assembly

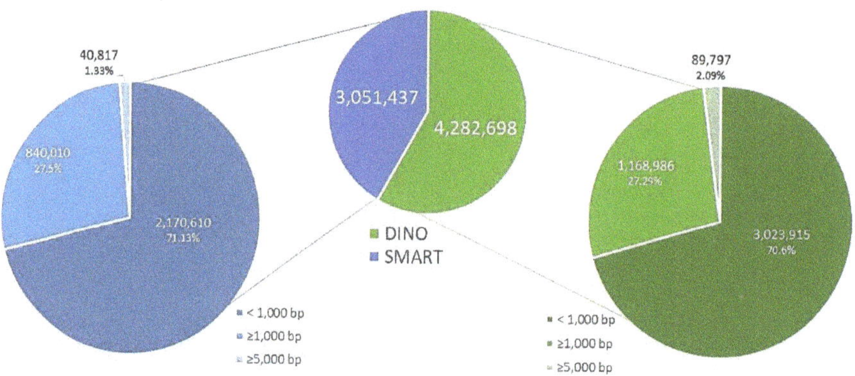

Figure 1. Number of sequences from the de novo assembly. The raw sequences were de novo assembled using Trinity (v2.8.4). The total number of sequences for the DINO (Dino Spliced Leader) library was 4,282,698; of these, 3,023,915 sequences were <1000 bp; 1,168,986 were from 1000 to 5000 bp; and 89,797 were >5000 bp. The total number of sequences for the SMART (Switching Mechanism at 5′ end of RNA Transcript) library was 3,051,437, distributed as follows: 2,170,610 were <1000 bp; 840,010 sequences between 1000 to 5000 bp; and 40,817 sequences >5000 bp.

3.3. Higher 5′ Representation in the DINO Library

In most known cDNA library databases, several sequences are truncated towards the 5′ end either partly or completely. This results in sequences lacking the initiation site and, therefore, upon further processing, they result in incorrect ORFs. Thus, in order to identify the translation initiation in unique sequences for each library, we analyzed the Kozak sequence for *Symbiodinium* RCCATGGCN described in Zhang et al. [11].

Among the 1422 unique genes in the DINO library, 1128 contained the Kozak sequence which represents 78.22% of the total. On the other hand, from the 1070 unique genes of the SMART library, 776 contained the Kozak sequence which represents 72.52% (Figure 2a).

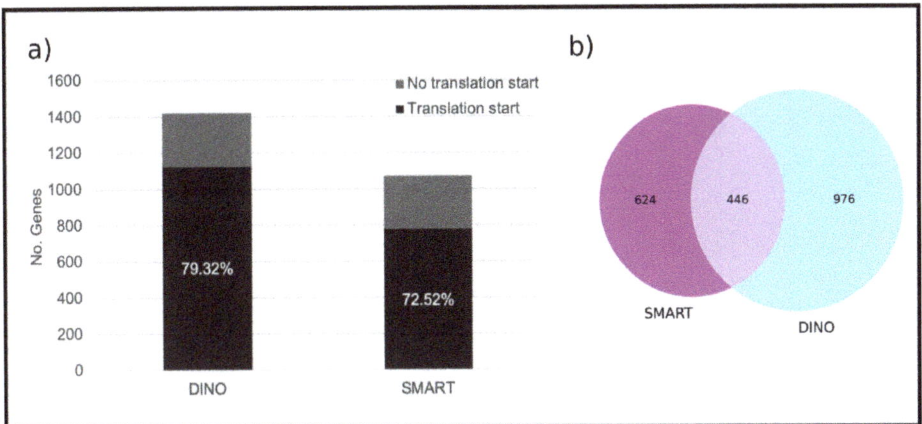

Figure 2. Total number of genes present in the libraries. (**a**) Percentage of sequences showing the Kozak sequence; (**b**) Venn diagram of the genes shared between both libraries.

3.4. Identification of Unique and Shared Orthologs in Assemblies

To identify the unique proteins in the libraries, the orthologous proteins between them, and those from the draft of the *S. microadriaticum* genome [22,23] were identified. A number of 1422 unique proteins was identified from the DINO library and 1070 unique proteins from the SMART library, respectively, including 446 proteins shared between the two libraries (Figure 2b). Upon functional annotation using KEGG for each of the libraries, the highest proportion of the proteins in the DINO library was found to be related to genetic information processing (256 genes), carbohydrate metabolism (65), signaling and cellular processes (37), and environmental information processing (35). In parallel, these proteins were found to be related to different metabolic pathways which included those involved in carbohydrate metabolism such as glycolysis/gluconeogenesis (13), glyoxylate and dicarboxylate metabolism (8), butanoate metabolism (7), the citric acid cycle, (6), pentose phosphate pathways (6) and those involved in energy metabolism such as photosynthesis (16), methane metabolism (10), and oxidative phosphorylation (8). Others were related to carbon fixation in photosynthetic organisms (8); lipid metabolism as fatty acid degradation (9), amino acid metabolism such as glycine, serine, and threonine metabolism (7), and cysteine and methionine metabolism (6), among others. On the other hand, the proteins in the SMART library were related to genetic information processing (179 genes), carbohydrate metabolism (72), signaling and cellular processes (37), and energy metabolism (30). Furthermore, these proteins were associated to different metabolic pathways which included carbohydrate metabolism as glycolysis/gluconeogenesis (12), starch and sucrose metabolism (9), the citrate cycle, (5), and pentose phosphate pathways (6); others were related to energy metabolism such as photosynthesis (14), carbon fixation in photosynthetic organisms (8), methane metabolism (6); and lipid metabolism such as fatty acid degradation (7); and finally, others related to amino acid metabolism such as valine, leucine and isoleucine degradation (8), and alanine, aspartate and glutamate metabolism (6), among others (Figure 3, Supplementary Table S1).

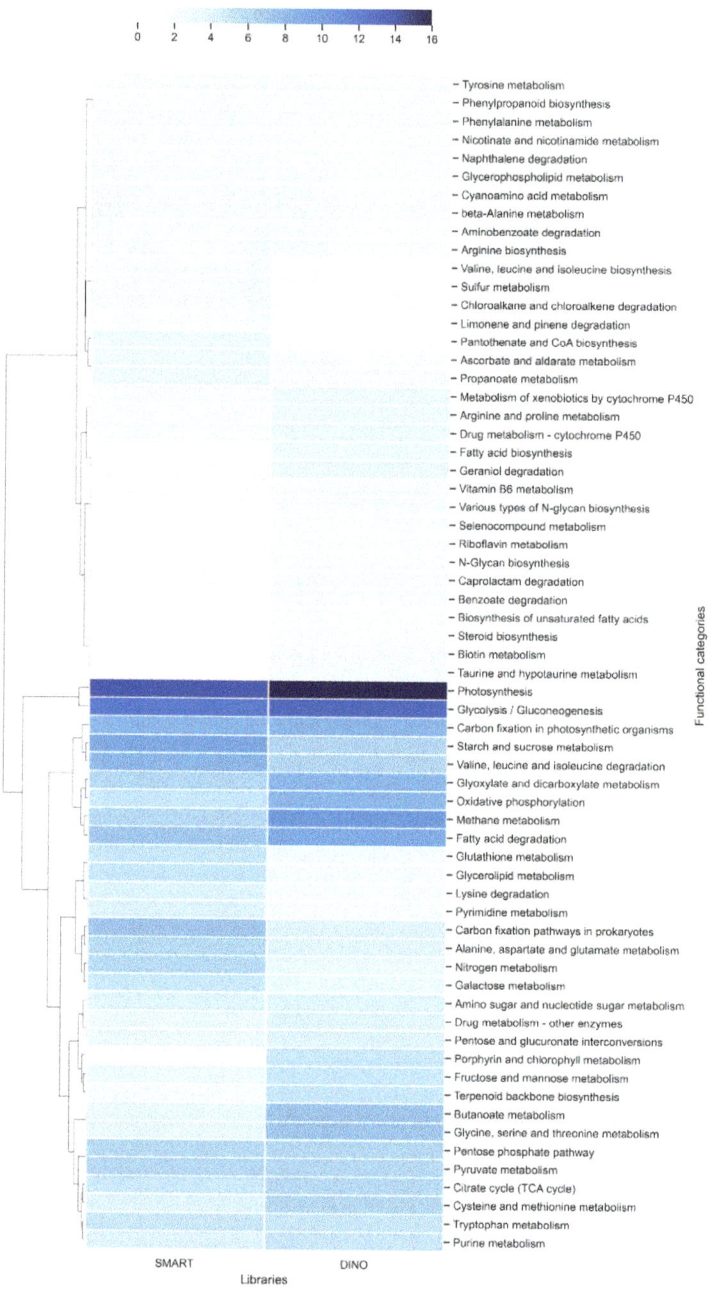

Figure 3. Heatmap of metabolic pathways associated with each library. A hierarchical cluster based on Euclidean distance measure and Ward's method for linkage analysis was carried out. Each row represents the metabolic pathways term.

3.5. Identification of the Receptor for Activated C Kinase 1 from *S. microadriaticum* (SmicRACK1) Ligands by Y2HS Using the DINO Library

One of the main concerns for using the Y2HS to identify protein-protein interactions is that with the use of heterologous systems like this one, the correct expression of heterologous proteins can be affected. Due to this, we made a Y2HS screen of putative SmicRACK1 ligands to prove that the DINO library can be suitable to identify protein-protein interactions. We sought to identify ligands of the Receptor for Activated C Kinase 1 from *S. microadriaticum* (SmicRACK1) [27]. Our Y2HS screen identified several putative SmicRACK1 ligands (Supplementary Figure S1) whose sequences, in general, contained the Dino-SL sequence, a short 5′ UTR and a 3′ UTR. One of these was identified as belonging to the hemerythrin-like superfamily (Figure 4) and presented again the Dino-SL sequence followed by a 5′ UTR (44 bp), a CDS with start and stop codons, and a 3′ UTR (Figure 4). A second Y2H assay using each sequence as bait and prey, correspondingly, resulted in the activation of the reporter gene and confirmed the interaction between these two proteins (Supplementary Figure S1). This indicated that the Dino-SL tag in dinoflagellates is a useful tool to generate cDNA libraries not only for sequencing, but also to identify ligands by Y2HS.

CCGTAGCCATTTTGGCTCAAGCCAGAGAGGCTCAATCCTAATTCGGTTCACCTCTCCAAGCTTCCATGGCTTTCGATT
GGCCAAACCCTTCGAACCCCAAGCCGACGTCCAAGGTGGTTACCGAAGCGAAGGACTGGCGGGACGAGGCGCTTCT
GGTGCCGCACGAGTGCATCCGCTGGTGGAATGCCAACCTATTGGAGATCCTGGCCGACTTCGAGCCGGTGAAGCGC
CCTACATCTGCGTGGAAGACGAAGGTACTTATGGATTTCATGGACAACTATTACGTTCCCTGTGTCCATCATCACCAC
CGCAGCGAGGAGGAAATCTACAATCCAGGCATCCTGGAAAAATGCAAGAGCATGGGCGTGGCTGATCCATTTACCA
AGGTGAAGAAGGATCACGAGACTCTCGTTTCTCTACTGGACAAGGTAGTCACGTTCCGGAAGCCGATCGAATCAGG
CGATGCCAAGGCAGTGGAGGAGTTCAAGACGGCCATGAAAGAGATGATCAAGTTCATGGAGGAGCACCTCGCCGA
GGAAGAGCAGGACTACCCCAAGATTTTCGCGGACTGTCAGCTTACGCAGGAGGAAGAGGGTGTTCTCTTGAACAAG
ATTTTGGAAGGCCTGGGCTTAGACGGTAGCAAGAAATTCTTGCCGCCCATCATGTACGCCATGGTCTTGTGGAGGGG
CAAGGACAAGATGATGGAGTGGGAGGCCAAGCTACCTGGGCCAACTCGCATGCTGAACAGCAACTGCTGGCTCAAC
GACTTCCACGAGAATCAACTCCGTGTACTGGAGGCTCTCAAGAAGGAAGAGGCATTCGAGCCTCACCCTCCGTCATG
CAATTTATGCTCCGTGATGTAACAAACTTGTCGCTGATCAGGTTGCAGTCGGAATTCTCTGGCGTCTTTTGGGTTGCG
AGATCTTTTTTTCTTGGCAAGA

DINO-SL 5′ and 3′ UTR Start codon Stop codon

Figure 4. The sequence of the putative Receptor for Activated C Kinase 1 from *S. microadriaticum* (SmicRACK1) ligand, hemerythrin-like protein identified by the Y2HS. The obtained sequence presents all the elements of a full-length transcript. The ligand sequence was identified after carrying out the Y2HS screening of the cDNA DINO library. The Dino-SL sequence is shown in yellow; the 5′ and 3′ UTR's in blue; the START and STOP codons are shown in green and red, respectively.

4. Discussion

The mechanisms underlying Symbiodiniaceae–cnidarian mutualistic symbiosis are still poorly understood even though it is a fundamental association for the survival and biodiversity of coral reefs. This association is also sensitive to environmental stress [28], particularly to the rise of temperature and light which causes its breakdown. Analogous to the former, there has been slow progress in the study of the molecular mechanisms involved in this disruption, mainly due to an absence of appropriate techniques to study the proteins, functions, and signaling pathways that work to maintain the symbiotic association. In order to develop a tool that could aid in the identification of molecular mechanisms that govern the life cycle and the Symbiodiniaceae–cnidarian symbiosis, we took advantage of the Dino-SL sequence found in the Symbiodiniaceae transcripts to generate a cDNA library from *S. microadriaticum* and analyze its performance and suitability in the Y2HS, compared to a SMART-generated library. In dinoflagellates, the spliced leader trans-splicing process transfers a Dino-SL sequence to the 5′ end of independent immature transcripts [7,8]. On the other hand, the SMART technology takes advantage of the reverse transcriptase ability to switch templates, reducing the occurrence of incomplete cDNA synthesis. However, the latter does not discriminate between full length and truncated transcripts [29], which results in a decrease of full-length cDNAs. The Y2HS is one of the molecular tools used to find in vivo novel protein–protein interactions by the reconstitution of a transcription factor that activates reporter genes. This approach has been used in assays with several cDNA libraries from different organisms, but not Symbiodiniaceae. Thus, we sought to explore the possible advantages and application of a DINO library from *S. microadriaticum* to identify putative SmicRACK1 ligands by Y2HS.

Our comparative analysis indicated that a cDNA library generated with the Dino-SL sequence was most suitable to obtain a higher recovery of transcript ORFs which, in addition, included more sequence of the 5′ end UTRs compared to that from the SMART library. In addition, the DINO library presented a higher number of independent clones (2.43×10^9) with respect to the SMART library (7.2×10^8) (Table 1). The number of independent clones is indicative of the library complexity, which implies that the DINO library had better representation of transcripts. Furthermore, the complexity of the libraries was confirmed by the total reads obtained after sequencing. Up to double total reads were obtained for the DINO library (35,311,434 M), compared with the SMART library (17,715,026 M) (Table 2). This improvement in the library yield results in an enrichment of transcripts without the use of expensive kits or complicated procedures and only an oligonucleotide with the Dino-SL sequence to replace the SMART technology is required. In practical terms for the Y2HS assay, this also results in sequences of the "prey" library without incorrect ORFs and capable of expressing complete proteins, thus elevating the probability of obtaining reporter-activated colonies through interacting ligands during the screening process.

It is important to point out that the sequences <1000 bp retrieved after the de novo assembly were present in more abundance in both libraries (70.6% and 71.13% for the DINO and SMART libraries, respectively) (Figure 1). This indicates the presence of truncated cDNAs in the two libraries. Likewise, the percent of sequences between 1000 and 5000 bp, and >5000 bp remained consistent in both libraries (Figure 1), which indicates that both methodologies have similar performance. Nevertheless, the libraries differ in the yield of recovered sequences because of the use of the Dino-SL instead of the SMART technology; the latter supposed to select for complete transcripts. The identified unique coding sequences were also higher for the DINO library (1422) than for the SMART library (1070); and the libraries shared only 446 of the coding sequences (Figure 2b). This low number of shared sequences is very likely the result of the SMART library harboring more incomplete CDS, therefore, escaping detection in the analyses. The number of coding sequences identified for the DINO library was lower than expected considering the high number of assembled sequences obtained; however, it is also a result of the multicopy genome of Symbiodiniaceae, whose genome presents multiple gene copies from gene duplication; i.e., the genome of *S. goreaui* displayed a genome-fragment duplication of 15.31% (5498 of 35,913) of the predicted genes [30]. This is also evident from their reported genomes, which have revealed a significant portion of genes arranged in tandem [31].

These data suggest that the increased number of sequences of the Dino-SL based library is likely due to a diminished loss of transcripts and 5′ ends, thus resulting in a better representation than in the SMART library. As the Dino-SL oligo selectively amplifies dT oligo-synthesized cDNA harboring the Dino-SL sequence, the DINO library should be enriched for sequences that contain the translation start site and upstream. It is well known that dinoflagellates such as *S. kawagutii*, use strong Kozak motifs as consensus sequences of the translation initiation site [11,32]. Thus, we examined the presence of the AUG start codon flanked by the Kozak sequence in the unique gene sequences of both libraries. There was a prevalence of a higher number of genes that presented this translation start site in the DINO library (78.22%, 1128 from 1422 genes) than those in the SMART library (72.52%, 776 from 1070 genes (Figure 2a). This indicates that by using the Dino-SL strategy, a greater coverage of the 5′ end, including the translation start site, which is an indispensable requirement for an expression library, is obtained.

The functional annotation of the libraries shows a high coincidence between both libraries; however, there were no proteins related to the categories metabolism of terpenoids and polyketides in the DINO library. On the other hand, there were no proteins related to the category biosynthesis of other secondary metabolites in the SMART library (Supplementary Table S1). In addition, the heatmap of metabolic pathways indicates a similar pattern in both libraries (Figure 3), but whose differences are the result of the difference in the methodologies (from generation of the libraries to sequencing). Thus, the differences in the methodologies did not have an important effect on the pool of transcripts or their abundance.

The spliced leader trans-splicing phenomenon is not particular for dinoflagellates. It was first described in trypanosomes [33–35], and later it was also identified in *Euglena* [36], nematodes [37], platyhelminthes [38], chordates [39–42], cnidarians [43], rotifers [44], dinoflagellates [7,45,46], porifers [47], ctenophors [47,48], and chaetognaths [49]. However, although there are several spliced leader consensus sequences in most of these organisms and the trans-splicing process does not occur in all the mRNAs, this feature is only useful in cases of a unique consensus sequence occurring in the whole mRNA population, i.e., rotifers, chordates and dinoflagellates [8].

SmicRACK1 belongs to the RACK1 and WD-40 protein families [27], it is expressed ubiquitously in eukaryotes, and is related to multiple cellular processes due to its scaffolding properties that enable it to interact with multiple ligands at the same time [50]. We conducted a Y2HS screen to identify SmicRACK1 ligands and to confirm the functionality of the DINO library. We identified several putative SmicRACK1 ligand sequences that contained the recognizable Dino-SL sequence, 5′ UTR, CDS and 3′ UTR. One of the ligand CDS encoded a Hemerythrin-like superfamily protein (Figure 4). Hemerythrin proteins belong to four family members of oxygen-carrier non-heme di-iron bonding proteins (hemoglobin, hemerythrin and two non-homologous families of arthropodan and molluscan hemocyanins) [51,52]. Hemerythrin proteins have been identified in invertebrates (sipunculans, polychaetes, priapulids and brachiopods) [53], and in 367 bacterial, 21 archaeal and 4 eukaryotic genomes [51]. Their functions are involved in signal transduction, phosphorelay regulation, abiotic resistance, and protein binding [51,52]. The function of this protein in *S. microadriaticum* is yet to be determined, but its characteristics point to an involvement in oxygen and iron storage and transport. Confirmation of its function will allow the physiological processes of *S. microadriaticum* in which SmicRACK1 participates to be identified.

5. Conclusions

Our comparative analysis clearly indicates that using the Dino-SL sequence found in the transcripts of *S. microadriaticum* to generate cDNA libraries allows a higher transcript recovery to be obtained when compared to a SMART technology-generated library. Furthermore, a combined strategy using both types of library will increase the probability of obtaining a more comprehensive coverage of transcripts to enhance the corresponding application to which the libraries are to be used. This molecular tool will enable progress in understanding the molecular events underlying the biology, metabolism and symbiosis of Symbiodiniaceae.

Supplementary Materials: The following are available online at https://www.mdpi.com/2076-260 7/9/4/791/s1, Figure S1: Y2HS positive clones indicating SmicRACK1 interacting ligands. Table S1: Functional annotation of libraries using GhostKOALA.

Author Contributions: Conceptualization, M.A.V. and T.I.-F.; methodology, T.I.-F.; software, E.G.-V.; validation, M.A.V., T.I.-F. and E.G.-V.; formal analysis, M.A.V., T.I.-F. and E.G.-V.; investigation, M.A.V.,

T.I.-F. and E.G.-V.; resources, M.A.V.; data curation, E.G.-V.; writing—original draft preparation, M.A.V., T.I.-F. and E.G.-V.; writing—review and editing, M.A.V., T.I.-F. and E.G.-V.; visualization, T.I.-F.; supervision, M.A.V.; project administration, M.A.V. and T.I.-F.; funding acquisition, M.A.V. All authors have read and agreed to the published version of the manuscript.

Funding: This work was supported by grants 285802 from CONACyT (National Council of Science and Technology-México) and IN-203718 from PAPIIT-UNAM.

Institutional Review Board Statement: Not applicable.

Informed Consent Statement: Not applicable.

Conflicts of Interest: The authors declare no conflict of interest.

References

1. Suggett, D.J.; Warner, M.E.; Leggat, W. Symbiotic Dinoflagellate Functional Diversity Mediates Coral Survival under Ecological Crisis. *Trends Ecol. Evol.* **2017**, *32*, 735–745. [CrossRef]
2. Teschima, M.M.; Garrido, A.; Paris, A.; Nunes, F.L.D.; Zilberberg, C. Biogeography of the endosymbiotic dinoflagellates (Symbiodiniaceae) community associated with the brooding coral Favia gravida in the Atlantic Ocean. *PLoS ONE* **2019**, *14*, e0213519. [CrossRef]
3. Ortiz-Matamoros, M.F.; Islas-Flores, T.; Voigt, B.; Menzel, D.; Baluška, F.; Villanueva, M.A. Heterologous DNA Uptake in Cultured Symbiodinium spp. Aided by Agrobacterium tumefaciens. *PLoS ONE* **2015**, *10*, e0132693. [CrossRef]
4. Ortiz-Matamoros, M.F.; Villanueva, M.A.; Islas-Flores, T. Genetic transformation of cell-walled plant and algae cells: Delivering DNA through the cell wall. *Brief. Funct. Genom.* **2017**, *17*, 26–33. [CrossRef]
5. Chow, M.H.; Yan, K.T.H.; Bennett, M.J.; Wong, J.T.Y. Birefringence and DNA Condensation of Liquid Crystalline Chromosomes. *Eukaryot. Cell* **2010**, *9*, 1577–1587. [CrossRef]
6. Nand, A.; Zhan, Y.; Salazar, O.R.; Aranda, M.; Voolstra, C.R.; Dekker, J. Chromosome-scale assembly of the coral endosymbiont Symbiodinium microadriaticum genome provides insight into the unique biology of dinoflagellate chromosomes. *bioRxiv* **2020**. [CrossRef]
7. Zhang, H.; Hou, Y.; Miranda, L.; Campbell, D.A.; Sturm, N.R.; Gaasterland, T.; Lin, S. Spliced leader RNA trans-splicing in dinoflagellates. *Proc. Natl. Acad. Sci. USA* **2007**, *104*, 4618–4623. [CrossRef]
8. Bitar, M.; Boroni, M.; Macedo, A.M.; Machado, C.R.; Franco, G.R. The spliced leader trans-splicing mechanism in different organisms: Molecular details and possible biological roles. *Front. Genet.* **2013**, *4*, 199. [CrossRef] [PubMed]
9. Zhang, H.; Lin, S. Retrieval of Missing Spliced Leader in Dinoflagellates. *PLoS ONE* **2009**, *4*, e4129. [CrossRef] [PubMed]
10. Lin, S.; Zhang, H.; Zhuang, Y.; Tran, B.; Gill, J. Spliced leader-based metatranscriptomic analyses lead to recognition of hidden genomic features in dinoflagellates. *Proc. Natl. Acad. Sci. USA* **2010**, *107*, 20033–20038. [CrossRef] [PubMed]
11. Zhang, H.; Zhuang, Y.; Gill, J.; Lin, S. Proof that Dinoflagellate Spliced Leader (DinoSL) is a Useful Hook for Fishing Dinoflagellate Transcripts from Mixed Microbial Samples: Symbiodinium kawagutii as a Case Study. *Protist* **2013**, *164*, 510–527. [CrossRef]
12. Song, B.; Chen, S.; Chen, W. Dinoflagellates, a Unique Lineage for Retrogene Research. *Front. Microbiol.* **2018**, *9*, 1556. [CrossRef]
13. Sugahara, Y.; Carninci, P.; Itoh, M.; Shibata, K.; Konno, H.; Endo, T.; Muramatsu, M.; Hayashizaki, Y. Comparative evaluation of $5'$-end-sequence quality of clones in CAP trapper and other full-length-cDNA libraries. *Gene* **2001**, *263*, 93–102. [CrossRef]
14. Agarwal, S.; Macfarlan, T.S.; Sartor, M.A.; Iwase, S. Sequencing of first-strand cDNA library reveals full-length transcriptomes. *Nat. Commun.* **2015**, *6*, 6002. [CrossRef] [PubMed]
15. Auerbach, D.; Stagljar, I. Yeast Two-Hybrid Protein-Protein Interaction Networks. In *Proteomics and Protein-Protein Interactions*; Metzler, J.B., Ed.; Springer: Boston, MA, USA, 2006; pp. 19–31.
16. Fields, S.; Song, O.-K. A novel genetic system to detect protein–protein interactions. *Nature* **1989**, *340*, 245–246. [CrossRef] [PubMed]
17. Van Criekinge, W.; Beyaert, R. Yeast two-hybrid: State of the art. *Biol. Proced. Online* **1999**, *2*, 1–38. [CrossRef] [PubMed]
18. Altschul, S.F.; Gish, W.; Miller, W.; Myers, E.W.; Lipman, D.J. Basic local alignment search tool. *J. Mol. Biol.* **1990**, *215*, 403–410. [CrossRef]
19. Blank, R.J. Cell architecture of the dinoflagellate Symbiodinium sp. inhabiting the Hawaiian stony coral Montipora verrucosa. *Mar. Biol.* **1987**, *94*, 143–155. [CrossRef]
20. Chen, S.; Zhou, Y.; Chen, Y.; Gu, J. Fastp: An ultra-fast all-in-one FASTQ preprocessor. *Bioinformatics* **2018**, *34*, i884–i890. [CrossRef]
21. Haas, B.J.; Papanicolaou, A.; Yassour, M.; Grabherr, M.; Blood, P.D.; Bowden, J.; Couger, M.B.; Eccles, D.; Li, B.; Lieber, M.; et al. De novo transcript sequence reconstruction from RNA-seq using the Trinity platform for reference generation and analysis. *Nat. Protoc.* **2013**, *8*, 1494–1512. [CrossRef]
22. Aranda, M.; Li, Y.; Liew, Y.J.; Baumgarten, S.; Simakov, O.; Wilson, M.C.; Piel, J.; Ashoor, H.; Bougouffa, S.; Bajic, V.B.; et al. Genomes of coral dinoflagellate symbionts highlight evolutionary adaptations conducive to a symbiotic lifestyle. *Sci. Rep.* **2016**, *6*, 39734. [CrossRef]
23. Liew, Y.J.; Aranda, M.; Voolstra, C.R. Reefgenomics. Org—A repository for marine genomics data. *Database* **2016**, *2016*. [CrossRef]

24. Wheeler, T.J.; Eddy, S.R. Nhmmer: DNA homology search with profile HMMs. *Bioinformatics* **2013**, *29*, 2487–2489. [CrossRef] [PubMed]

25. Kanehisa, M.; Sato, Y.; Morishima, K. BlastKOALA and GhostKOALA: KEGG Tools for Functional Characterization of Genome and Metagenome Sequences. *J. Mol. Biol.* **2016**, *428*, 726–731. [CrossRef] [PubMed]

26. Nguyen, N.T.T.; Contreras-Moreira, B.; Castro-Mondragon, J.A.; Santana-Garcia, W.; Ossio, R.; Robles-Espinoza, C.D.; Bahin, M.; Collombet, S.; Vincens, P.; Thieffry, D.; et al. RSAT 2018: Regulatory sequence analysis tools 20th anniversary. *Nucleic Acids Res.* **2018**, *46*, W209–W214. [CrossRef] [PubMed]

27. Islas-Flores, T.; Pérez-Cervantes, E.; Nava-Galeana, J.; Loredo-Guillén, M.; Guillén, G.; Villanueva, M.A. Molecular Features and mRNA Expression of the Receptor for Activated C Kinase 1 from Symbiodinium microadriaticum ssp. microadriaticum During Growth and the Light/Dark cycle. *J. Eukaryot. Microbiol.* **2018**, *66*, 254–266. [CrossRef] [PubMed]

28. Muscatine, L.; Porter, J.W. Reef Corals: Mutualistic Symbioses Adapted to Nutrient-Poor Environments. *Bioscience* **1977**, *27*, 454–460. [CrossRef]

29. Cartolano, M.; Huettel, B.; Hartwig, B.; Reinhardt, R.; Schneeberger, K. cDNA Library Enrichment of Full Length Transcripts for SMRT Long Read Sequencing. *PLoS ONE* **2016**, *11*, e0157779. [CrossRef]

30. Liu, H.; Stephens, T.G.; González-Pech, R.A.; Beltran, V.H.; Lapeyre, B.; Bongaerts, P.; Cooke, I.; Aranda, M.; Bourne, D.G.; Forêt, S.; et al. Symbiodinium genomes reveal adaptive evolution of functions related to coral-dinoflagellate symbiosis. *Commun. Biol.* **2018**, *1*, 95. [CrossRef]

31. Shoguchi, E.; Shinzato, C.; Kawashima, T.; Gyoja, F.; Mungpakdee, S.; Koyanagi, R.; Takeuchi, T.; Hisata, K.; Tanaka, M.; Fujiwara, M.; et al. Draft Assembly of the Symbiodinium minutum Nuclear Genome Reveals Dinoflagellate Gene Structure. *Curr. Biol.* **2013**, *23*, 1399–1408. [CrossRef]

32. Bodył, A.; Mackiewicz, P. Analysis of the targeting sequences of an iron-containing superoxide dismutase (SOD) of the dinoflagellate Lingulodinium polyedrum suggests function in multiple cellular compartments. *Arch. Microbiol.* **2006**, *187*, 281–296. [CrossRef]

33. Sather, S.; Agabian, N. A 5′ spliced leader is added in trans to both alpha- and beta-tubulin transcripts in Trypanosoma brucei. *Proc. Natl. Acad. Sci. USA* **1985**, *82*, 5695–5699. [CrossRef]

34. Murphy, W.J.; Watkins, K.P.; Agabian, N. Identification of a novel Y branch structure as an intermediate in trypanosome mRNA processing: Evidence for Trans splicing. *Cell* **1986**, *47*, 517–525. [CrossRef]

35. Sutton, R.E.; Boothroyd, J.C. Evidence for Trans splicing in trypanosomes. *Cell* **1986**, *47*, 527–535. [CrossRef]

36. Tessier, L.; Keller, M.; Chan, R.; Fournier, R.; Weil, J.; Imbault, P. Short leader sequences may be transferred from small RNAs to pre-mature mRNAs by trans-splicing in Euglena. *EMBO J.* **1991**, *10*, 2621–2625. [CrossRef]

37. Krause, M.; Hirsh, D. A trans-spliced leader sequence on actin mRNA in C. elegans. *Cell* **1987**, *49*, 753–761. [CrossRef]

38. Rajkovic, A.; Davis, R.E.; Simonsen, J.N.; Rottman, F.M. A spliced leader is present on a subset of mRNAs from the human parasite Schistosoma mansoni. *Proc. Natl. Acad. Sci. USA* **1990**, *87*, 8879–8883. [CrossRef] [PubMed]

39. Vandenberghe, A.E.; Meedel, T.H.; Hastings, K.E. mRNA 5′-leader trans-splicing in the chordates. *Genes Dev.* **2001**, *15*, 294–303. [CrossRef] [PubMed]

40. Yuasa, H.J.; Kawamura, K.; Yamamoto, H. The Structural Organization of Ascidian Halocynthia roretzi Troponin I Genes. *J. Biochem.* **2002**, *132*, 135–141. [CrossRef] [PubMed]

41. Ganot, P.; Kallesøe, T.; Reinhardt, R.; Chourrout, D.; Thompson, E.M. Spliced-Leader RNA trans Splicing in a Chordate, Oikopleura dioica, with a Compact Genome. *Mol. Cell. Biol.* **2004**, *24*, 7795–7805. [CrossRef]

42. Satou, Y. Genomic overview of mRNA 5′-leader trans-splicing in the ascidian Ciona intestinalis. *Nucleic Acids Res.* **2006**, *34*, 3378–3388. [CrossRef]

43. Stover, N.A.; Steele, R.E. Trans-spliced leader addition to mRNAs in a cnidarian. *Proc. Natl. Acad. Sci. USA* **2001**, *98*, 5693–5698. [CrossRef]

44. Pouchkina-Stantcheva, N.N.; Tunnacliffe, A. Spliced Leader RNA–Mediated trans-Splicing in Phylum Rotifera. *Mol. Biol. Evol.* **2005**, *22*, 1482–1489. [CrossRef]

45. Lidie, K.B.; Van Dolah, F.M. Spliced Leader RNA-Mediated trans-Splicing in a Dinoflagellate, Karenia brevis. *J. Eukaryot. Microbiol.* **2007**, *54*, 427–435. [CrossRef]

46. Bachvaroff, T.R.; Place, A.R. From Stop to Start: Tandem Gene Arrangement, Copy Number and Trans-Splicing Sites in the Dinoflagellate Amphidinium carterae. *PLoS ONE* **2008**, *3*, e2929. [CrossRef]

47. Douris, V.; Telford, M.J.; Averof, M. Evidence for Multiple Independent Origins of trans-Splicing in Metazoa. *Mol. Biol. Evol.* **2009**, *27*, 684–693. [CrossRef]

48. Derelle, R.; Momose, T.; Manuel, M.; Da Silva, C.; Wincker, P.; Houliston, E. Convergent origins and rapid evolution of spliced leader trans-splicing in Metazoa: Insights from the Ctenophora and Hydrozoa. *RNA* **2010**, *16*, 696–707. [CrossRef] [PubMed]

49. Marlétaz, F.; Gilles, A.; Caubit, X.; Perez, Y.; Dossat, C.; Samain, S.; Gyapay, G.; Wincker, P.; Le Parco, Y. Chætognath transcriptome reveals ancestral and unique features among bilaterians. *Genome Biol.* **2008**, *9*, R94. [CrossRef] [PubMed]

50. Islas-Flores, T.; Rahman, A.; Ullah, H.; Villanueva, M.A. The Receptor for Activated C Kinase in Plant Signaling: Tale of a Promiscuous Little Molecule. *Front. Plant Sci.* **2015**, *6*, 1090. [CrossRef] [PubMed]

51. Alvarez-Carreño, C.; Becerra, A.; Lazcano, A. Molecular Evolution of the Oxygen-Binding Hemerythrin Domain. *PLoS ONE* **2016**, *11*, e0157904. [CrossRef] [PubMed]

52. Li, X.; Li, J.; Hu, X.; Huang, L.; Xiao, J.; Chan, J.; Mi, K. Differential roles of the hemerythrin-like proteins of Mycobacterium smegmatis in hydrogen peroxide and erythromycin susceptibility. *Sci. Rep.* **2015**, *5*, 16130. [CrossRef] [PubMed]
53. Klotz, I.M.; Klippenstein, G.L.; Hendrickson, W.A. Hemerythrin: Alternative oxygen carrier. *Science* **1976**, *192*, 335–344. [CrossRef] [PubMed]

 *microorganisms*

Communication

Response of Coral Reef Dinoflagellates to Nanoplastics under Experimental Conditions Suggests Downregulation of Cellular Metabolism

Christina Ripken [1,2,*] , Konstantin Khalturin [2] and Eiichi Shoguchi [2]

1. Light-Matter Interactions for Quantum Technologies Unit, Okinawa Institute of Science and Technology Graduate University, Onna, Okinawa 904-0495, Japan
2. Marine Genomics Unit, Okinawa Institute of Science and Technology Graduate University, Onna, Okinawa 904-0495, Japan; konstantin.khalturin@oist.jp (K.K.); eiichi@oist.jp (E.S.)
* Correspondence: christina.ripken@oist.jp

Received: 25 September 2020; Accepted: 5 November 2020; Published: 9 November 2020

Abstract: Plastic products contribute heavily to anthropogenic pollution of the oceans. Small plastic particles in the microscale and nanoscale ranges have been found in all marine ecosystems, but little is known about their effects upon marine organisms. In this study, we examine changes in cell growth, aggregation, and gene expression of two symbiotic dinoflagellates of the family Symbiodiniaceae, *Symbiodinium tridacnidorum* (clade A3), and *Cladocopium* sp. (clade C) under exposure to 42-nm polystyrene beads. In laboratory experiments, the cell number and aggregation were reduced after 10 days of nanoplastic exposure at 0.01, 0.1, and 10 mg/L concentrations, but no clear correlation with plastic concentration was observed. Genes involved in dynein motor function were upregulated when compared to control conditions, while genes related to photosynthesis, mitosis, and intracellular degradation were downregulated. Overall, nanoplastic exposure led to more genes being downregulated than upregulated and the number of genes with altered expression was larger in *Cladocopium* sp. than in *S. tridacnidorum*, suggesting different sensitivity to nano-plastics between species. Our data show that nano-plastic inhibits growth and alters aggregation properties of microalgae, which may negatively affect the uptake of these indispensable symbionts by coral reef organisms.

Keywords: nanoplastics; dinoflagellate; coral reef; *Symbiodinium*; *Cladocopium*; gene expression

1. Introduction

Coral reefs provide a habitat for marine invertebrate and vertebrate species alike, sustaining the highest biodiversity among marine ecosystems [1]. Formed primarily by scleractinian corals and coralline algae, coral reefs are complex and vulnerable ecosystems. Structural complexity of coral reefs, and, by extension, the capability to sustain biodiversity often declines due to natural and human-related stressors [2,3].

One important stressor for coral reef ecosystems is plastic pollution. Small plastic particles (>1 mm) have been reported from coral islands at more than 1000 items/m^2 [4]. Further fragmentation of these particles leads to nano-plastics (<1 μm) [5]. Microplastic particles induce stress responses in scleractinian corals, and suppress their immune systems and capacity to cope with environmental toxins [6]. When ingested by corals [7–9], microplastics disrupt the anthozoan-algal symbiotic relationship [10]. They are also linked to potential adverse effects on calcification [11] with exposure resulting in attachment of microplastic particles to tentacles or mesenterial filaments, ingestion of microplastic particles, and increased mucus production [12]. Su et al. [13] exposed the coral symbiont, *Cladocopium goreaui*, to 1-μm polystyrene spheres, leading to diminished detoxification activity, nutrient uptake, and photosynthesis as well as increased oxidative stress, apoptosis levels, and ion transport. Plastic particles seem to negatively impact symbiotic relationships between corals and their microalgae, thereby degrading the entire coral reef ecosystem. However, this has not been systematically investigated.

Nano-plastics can originate by fragmentation of larger plastic objects through photochemical and mechanical degradation. There are also primary sources of nano-plastics. Medical and cosmetic products, nanofibers from clothes and carpets, 3D printing, and Styrofoam byproducts find their way into coral reef ecosystems via river drainages, sewage outfalls, and runoff after heavy rainfall, as well as via atmospheric input and ocean currents. Nano-plastics have recently been reported in ocean surface water samples [14]. Since the nanoplastic research is still in its infancy, many unanswered questions remain, starting with the environmental concentrations in various ecosystems [15,16]. Since detecting nano-plastics' concentrations directly is still not possible [17], a better understanding of the potential impacts is necessary to encompass a range of different concentrations. The miniature size of these particles leads to higher surface area to volume ratios and enhanced reactivity of smaller particles coupled with the ability to pass across biological barries and enter cells [18] when compared to micro-plastics.

In this study, we focused on the microalgal symbionts of mollusks that inhabit fringing coral reefs of Okinawa. Knowledge of the effects of nano-plastics on the symbionts of Tridacninae (giant clams) and Fraginae (heart cockles) will benefit conservation and restocking efforts, since both are obligatory photo-symbionts and important contributors to coral reef ecosystems. Approximately 30 Symbiodiniaceae phylotypes are economically important for fisheries [19]. This study specifically investigated effects of nano-plastics (42-nm polystyrene spheres) on the growth rates, aggregations, and gene expression changes in *Symbiodinium tridacnidorum* (symbionts of the Tridacninae) and *Cladocopium* sp. (symbionts of the Fraginae).

2. Materials and Methods

2.1. Exposure to Nano-Plastics Using Roller Tanks

The majority of host animals obtain their indispensable symbiotic dinoflagellates from coral reef sand and the water column [20,21]. Roller tanks and tables were used to simulate the natural environment of the dinoflagellate vegetative cells in their free-living state [22,23]. Roller tanks have commonly been used to promote aggregation since Shanks and Edmondson [23,24]. Fifteen roller tanks of 13.4 cm in diameter and 7.5 cm in height with a capacity of 1057 mL were employed. In tanks, aggregation can occur [23], ensuring that microalgae are exposed to the polystyrene nano-plastics (nanoPS) in a way that mimics their natural habitat. Once rotation commenced, continuous aggregate formation and suspension were ensured [24] as well as continuous exposure to nanoPS. Roller tanks are closed for the entire duration of the experiment, so that exposure levels of the nanoPS remain constant throughout. Tanks were closed without bubbles so as not to disturb the aggregation process with turbulence. To compare differences between species, two dinoflagellates, *Symbiodinium tridacnidorum* (clade A3 strain, ID: NIES-4076) and *Cladocopium* sp. (clade C strain, ID: NIES-4077) were cultured in artificial seawater containing 0.2× Guillard's (F/2) marine-water enrichment solution (Sigma-Aldrich) in roller tanks [25,26]. *S. tridacnidorum* and *Cladocopium* sp. were isolated from *Tridacna crocea* and *Fragum* sp. in Okinawa, Japan [25]. Using glass flasks, precultures for the stress experiment were established, as previously described [26].

Microplastics (>1 mm) from a coral reef and the ingestion (53 to 500 μm) by coral reef clams have been reported and microplastic removal by giant clams has been proposed [4,27]. To simulate nano-plastic accumulation in coral reefs and in the host organisms, three different concentrations (0.01 mg/L, 0.1 mg/L, and 10 mg/L) of nano-plastic (42-nm pristine polystyrene beads, nanoPS$_{42}$, from Bangs Laboratories Inc., catalog number FSDG001, polystyrene density 1.05 g/cm^3, nanoPS) were added to the treatment tanks (Table S1). Preliminary tests were run to confirm no leaching of the fluorescent dye (data not shown). Concentrations were chosen to span a range of possible environmental concentrations, starting at 0.01 mg/L with a surface area of 1.36×10^6 μm^2/L and 2.46×10^8 particles per L. The next highest concentration is just one magnitude higher (0.1 mg/L, surface area 1.36×10^7 μm^2/L and 2.46×10^9 particles per L). This middle range concentration corresponds to actually observed lower concentrations of microplastic particles [28]. Just as microplastic concentrations are highly variable, nanoplastic concentrations are assumed to change depending on the proximity to human activity. To account for these variables, but not at the highest measured microplastic concertation, we placed our highest concentration at 10 mg/L with a surface area of 1.36×10^9 μm^2/L and 2.46×10^{11} particles per L (Table S1). Treatment tanks as well as control tanks (no nanoPS) were established in triplicate. Three tanks without algae were prepared as negative controls (at 10 mg/L, 0.01 mg/L, and 0 mg/L nano-plastic). In each culture tank, the final cell density of the two strains was adjusted to ~7×10^5 cells/mL. Tanks were harvested after 9–11 days for logistical reasons, making replicates a day apart (Table S2).

2.2. Measurements of Cell Density and Aggregation

Cells for growth rates were counted using hemocytometers (C-Chip DHC-N01) under a Zeiss Axio Imager Z1 microscope (Jena, Germany). At least two subsamples and 200 cells were counted per sample.

Aggregates were imaged and counted in each tank and for five size classes, as follows: 0.2–0.5 mm, 0.5–1 mm, 1–2.5 mm, 2.5–3.5 mm, and >3.5 mm in the longest dimension. Tanks of the same concentration were sampled at the same time of day. Controls were sampled first and then in order of increasing nanoPS$_{42}$ concentration to avoid nano-plastic carry over from higher concentrations to lower. In order to examine how nanoPS$_{42}$ affects aggregate formation, aggregates were collected for different measurements after the approximate total number of aggregates in each tank had been determined. Aggregation of algae and plastic was confirmed with 3D imaging using a Zeiss Lightsheet Z.1 and Imaris software. NanoPS$_{42}$ was observed with a band-pass filter (excitation: 405 nm, emission: 505–545 nm) and chloroplasts were visualized using a long-pass red filter (excitation: 488 nm, emission: 660 nm).

One fourth of all aggregates were collected for RNA analysis (2 min spin down at 12,000 rpm and discarding the supernatant, freezing in liquid nitrogen, and storage at $-80\ ^\circ$C). For all other measured factors, harvest included separate sampling of the aggregate fraction (aggregates >0.5 mm, Agg) and the surrounding sea water fraction (aggregates <0.5 mm and un-aggregated cells) [29]. Aggregates for sinking velocity (three aggregates per size class for 11.5 cm in a 100-mL glass graduated glassware cylinder) was collected in artificial seawater at the same temperature as experiments were conducted.

2.3. RNA Extraction, Library Construction, and Sequencing

Frozen cells were broken mechanically using a polytron (KINEMATICA Inc., Luzern, Switzerland) in tubes chilled with liquid nitrogen. RNAs were extracted using Trizol reagent (Invitrogen) according to the manufacturer's protocol. The quantity and quality of total RNA were checked using a Qubit fluorometer (ThermoFisher, Waltham, MA) and a TapeStation (Agilent, Santa Clara, CA), respectively. Libraries for RNA-seq were constructed using the NEBNext Ultra II Directional RNA Library Prep Kit for Illumina (#E7760, NEB). Sequencing was performed on a NovaSeq6000 SP platform. Nine mRNA-seq libraries from nanoPS-exposed photosymbiotic algae were sequenced (3 concentrations × 3 exposure times) plus three controls (Table S2).

2.4. RNA-Seq Data Mapping and Clustering Analysis

Raw sequencing data obtained from the NovaSeq6000 were quality trimmed with Trimmomatic (v. 0.32) in order to remove adapter sequences and low-quality reads. Paired reads that survived the trimming step (on average 92%) were mapped against reference transcripts of *Symbiodinium* and *Cladocopium* sp. For each gene in the genomes of *Symbiodinium* and *Cladocopium* sp. a *.t1 transcript form was used as a reference sequence. Mapping was performed using RSEM (RNA-Seq by Expectation-Maximization) [30] with the bowtie (v. 1.1.2) as an alignment tool. Expression values across all samples were normalized by the TMM (Trimmed Mean of *M*-values) method [31]. Genes with differential expression (two-fold difference and $p < 0.001$) were identified with edgeR Bioconductor, based on the matrix of TMM normalized TPM Transcripts Per Kilobase) values. Experimental samples were clustered according to their gene expression characteristics using edgeR. Annotations were performed using BLAST2GO and Pfam databases [25] and are available at the genome browser site (https://marinegenomics.oist.jp).

3. Results and Discussion

3.1. Suppression of Algal Growth by Nano-Plastic Exposure

Exposure to nanoPS$_{42}$ decreased the mean growth rate of photosymbiotic algae (Figures 1 and S1). The greatest reduction in growth rate was seen at the lowest nanoPS$_{42}$ treatment (0.01 mg/L) with cell densities reduced from starting values by -0.062 ± 0.02 (Holm-Sidak, $p = 0.002$), which was followed by the highest nanoPS$_{42}$ treatment (10 mg/L) with -0.013 ± 0.05 (Holm-Sidak, $p = 0.026$). In the 0.1 mg/L treatment, cell densities increased slightly by 0.028 ± 0.04. Thus, nanoPS$_{42}$ either inhibited algal growth in a non-linear manner or had a limited effect [32]. Reductions in growth rates have also been reported in the micro-plastic study of Reference [13] in *Cladocopium goreaui* and in other microalgae exposed to micro-plastics (*Chlamydomonas reinhardtii* [33] and *Skeletonema costatum* [34]).

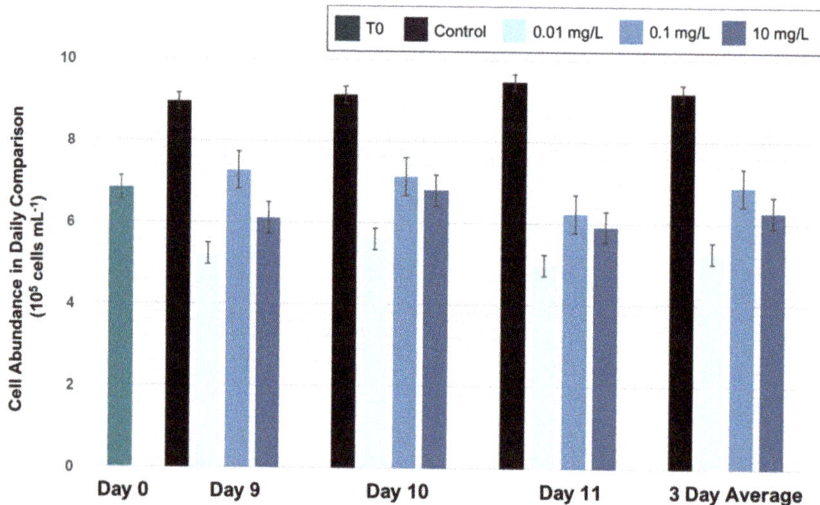

Figure 1. Treatment and control tanks were sampled after 9, 10, and 11 days. Experiments started with ~680,000 cells/mL in all tanks. There are differences between the growth rate in the different treatments, but the ratio stays the same over all three sampling days. The cell density in the control was $9.83 \pm 0.39 \times 10^5$ cells per mL, while treatment tanks were significantly lower: 0.01 mg/mL: $5.69 \pm 0.12 \times 10^5$ cells per mL, 0.1 mg/mL: $7.51 \pm 0.34 \times 10^5$ cells per mL, and 10 mg/mL: $6.96 \pm 0.40 \times 10^5$ cells per mL. Bars display a confidence interval.

In addition, Su et al. [13] reported a reduction in cell size in *Cladocopium goreaui*. Further investigations are needed to see if this is the case under nano-plastic exposure. The biggest growth rate reduction observed was at 0.01 mg/L nanoPS$_{42}$, which is far below the 5 mg/L used by Su et al. [13]. The nutrient deficiency is also a reason discussed in Reference [23], which could explain the larger effects on growth rates at lower concentrations. The reason for nutrient limitation induced by plastic is proposed to be interactions of the nutrients with the surface of the plastics [35]. NanoPS$_{42}$ self-aggregation could account for the higher nanoPS$_{42}$ treatments having less of an effect on the growth rates.

3.2. Nano-Plastic Exposure Influences the Number and Sinking Velocity of Cell Aggregates

To understand the impact of nanoPS$_{42}$ on aggregation in these two Symbiodiniaceae cultures, the total number of algal aggregates per tank and in five aggregate class sizes was recorded (Figures 2 and S2). All tanks showed aggregation, which was expected, as self-aggregation of Symbiodiniaceae has been observed previously [13].

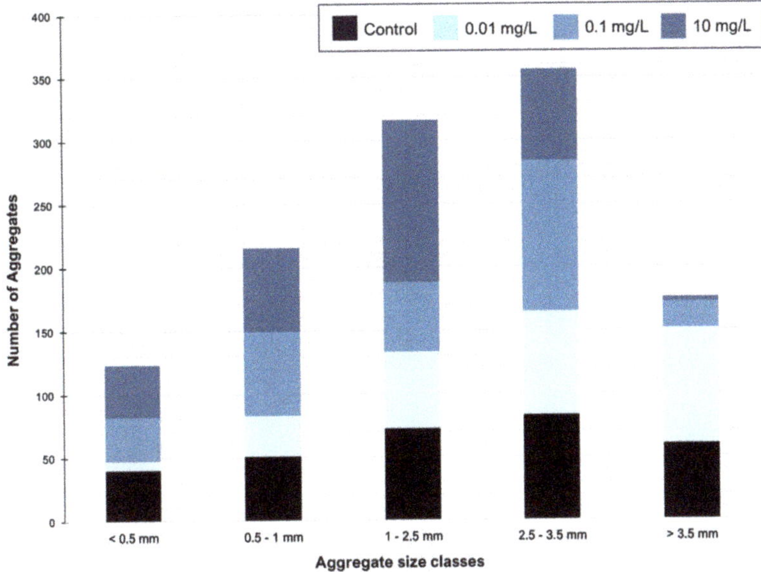

Figure 2. NanoPS exposure leads to a change in aggregation. Aggregates sorted by class size show a significant change in the distribution pattern under nanoPS$_{42}$ exposure (Holm-Sidak, $p = 0.05$). No differences are observed when exposure length is compared.

The majority of aggregates exhibited an ovoid form. A significant difference can be observed when aggregate numbers are compared over all class sizes and all treatments, showing that the nanoPS has an influence on the aggregation process. The lowest nanoPS$_{42}$ treatments (0.01 mg/L) shows significant reduction in the total aggregates count by 10% (Holm-Sidak, $p = 0.003$), but aggregation was enhanced overall in that treatment to have a higher percentage of huge aggregates than in the control treatment (Holm-Sidak, $p = 0.001$). While there is also a reduction of 3% in the intermediate nanoPS$_{42}$ treatment (0.1 mg/L), this is not significant (Holm-Sidak, $p = 0.319$). In the highest plastic treatment at 10 mg/L, this is reversed, leading to more aggregates overall, and more of those being of smaller sizes. The different aggregate class sizes show significantly different distributions in all three treatments and the control (ANOVA, $p < 0.001$) (Figure S2). In the control, the self-aggregation led to a specific distribution pattern of aggregate sizes, which was not repeated in the treatments. Self-aggregation was also observed in the microplastic experiments of Su et al. [13]. The fact that the presence of nanoPS changes the aggregation between the cells and leads to more aggregates in the bigger size classes is possible due to higher production of extracellular polymeric substances (EPS) with sticky properties, trapping more cells in one aggregate and keeping aggregates closer together.

Nutrient depletion, which has been linked to the presence of micro-plastics in algae cultures [35], is associated with increased stickiness of the EPS [36,37]. Differences in the EPS production due to the presence of nanoPS is a likely factor contributing to the differences in aggregation seen in the study. EPS production was not measured, so further studies are needed to confirm this hypothesis linking the aggregation process and EPS production in Symbiodiniaceae under nanoPS influence. Lagarde et al. [33] notices different aggregate formation under different plastic treatment and sizes, which matches with our results. In addition, in *Symbiodinium tridacnidorum*, genes encoding a protein with a TIG (Transcription factor immunoglobulin) domain were upregulated. Since this protein is found in surface cell receptors, it may influence changes in hetero aggregation.

Significant differences are evident when aggregate numbers are compared over size classes and treatments, showing that nanoPS influences aggregation. Aggregate size classes show significantly different distributions in all three treatments vs. controls (ANOVA, $p < 0.001$) (see Figure 2). These differences in aggregation could be due to changes of the cell surface receptors, as nanoPS increases genes related to those two-fold (see Section 3.3. NanoPS effects on gene expression).

Due to nanoPS exposure, aggregation and sinking velocities are impacted, which, in turn, leads to change in sedimentation. Since the majority of the host animals obtain their symbiotic dinoflagellates from the sand and water column [20], these changes in dinoflagellate sedimentation might lead to problems in acquisition of symbionts for the host animals. The lowest plastic treatment used, which is environmentally possible, already induces changes to the sedimentation. This lowest treatment led to bigger aggregates, which, at the same time, sank faster, possibly removing the symbionts from the water column faster than required from the host animals and reducing chances of encountering symbionts.

Changes in aggregation and resulting sedimentation were observed under nanoPS exposure (Figure 3). The biggest changes in sinking velocity correspond to increases in aggregation and are observed in the lowest plastic treatment at 0.01 mg/L. On the other hand, the 10 mg/L treatment did not have any significant effect on the sinking rates but did affect sedimentation indirectly through changes in the aggregate size distribution (Figure 3). These changes, including both sinking velocities and aggregate size distribution, are most likely due to hetero-aggregation between algae and nanoPS. Under different treatments, the size distribution of aggregates was significantly different (Figure 2). In combination, it is likely that the same effect that led to that difference in aggregation is also responsible for the difference in sinking velocities. Changes in EPS production and stickiness will lead to different cell packaging within the aggregates, possibly creating tighter packed aggregates in the lowest and intermediate treatment. This effect might be counteracted under the highest nanoPS exposure by the sheer volume of EPS, which is lighter than seawater. The nano-plastic itself trapped in these could also add to the sinking velocity returning back to control levels in the high plastic treatments. Since these symbionts are paired with the mobile larvae of the host animals, a higher sinking velocity would remove the potential symbiont from the pelagic area and reduce the chance of a match.

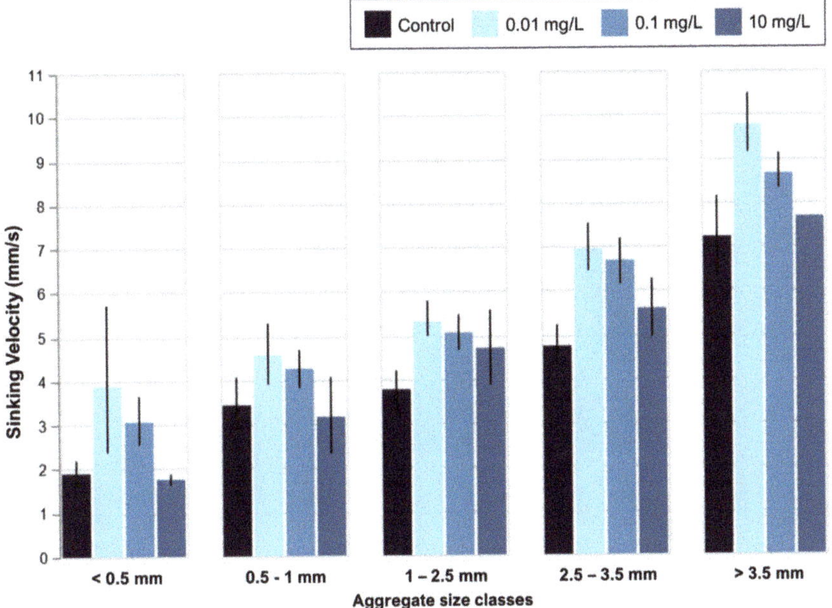

Figure 3. Sinking velocity change with nanoPS exposure. Sinking velocities decrease with aggregate size from more than 7 mm/s (>3.5 mm) to less than 2 mm/s (<0.5 mm). In all class sizes, the control was similar in sinking velocity to the highest nanoPS treatment (10 mg/L). The low nanoPS treatment (0.01 mg/L) differed significantly from both controls (*t*-test, two-tailed $p = 5.56 \times 10^{-4}$) and the highest nanoPS treatment (*t*-test, two-tailed $p = 9.03 \times 10^{-4}$). This was also true for the intermediate nanoPS treatment (darker blue, 0.1 mg/L). Error bars are 95% confidence intervals. Only one huge aggregate was measured in the highest nanoPS treatment. No differences in sinking velocity were observed in relation to exposure length.

3.3. NanoPS Effects on Gene Expression

Analysis of differential gene expression showed that, in *Symbiodinium*, 14 genes were upregulated after nanoPS$_{42}$ exposure, and 34 were downregulated relative to controls (Figure 4a). In *Cladocopium*, 75 genes were upregulated, and 169 genes were downregulated (Figure 4b). *Cladocopium* seems more sensitive to nanoPS$_{42}$ exposure, as overall more genes responded than in *Symbiodinium*. Since Pfam analysis had more annotations than BLAST2GO in differentially expressed genes (DEGs) of *Cladocopium*, we list the major domains encoded by the DEGs of *Cladocopium* (Tables S3–S6).

The largest group of upregulated genes was a subfamily of dynein-related proteins having an AAA_5 domain (Table 1). Dynein is a microtubule-associated motor protein. Ten genes for dynein-related proteins with AAA and/or DHC (Dynein heavy chain) were upregulated in *Cladocopium* sp. by nanoPS$_{42}$ (Table 1 and Table S4). It has been shown that microplastic exposure induces production of reactive oxygen species (ROS) in microalgae [13,33] and dynein upregulation. Therefore, it might be needed to balance cytoskeletal dynamics as microtubule polymerization is impaired by oxidative stress [38]. Dynein light chain genes were also shown to be upregulated in gill cells of zebra mussels exposed to polystyrene micro-plastic [39].

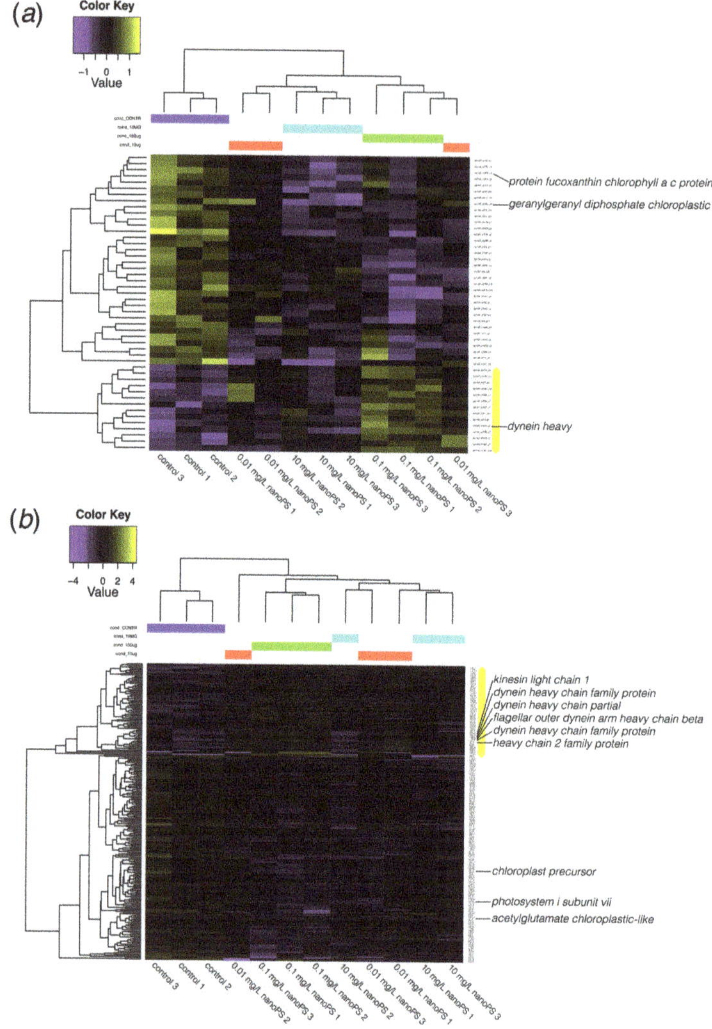

Figure 4. Heatmap and clustering of differentially expressed genes (2-fold changes, $p < 0.001$) between dinoflagellates exposed to nano-plastics and controls. (**a**) DEGs in *Symbiodinium tridacnidorum*. (**b**) DEGs in *Cladocopium* sp. Values indicate the relative gene expression level with purple and yellow showing downregulation and upregulation, respectively. The yellow bar shows a cluster of upregulated genes. Annotations by Blast2GO show the presence of microtubule-related or photosynthesis-related genes among DEGs.

Table 1. Domains encoded by more than three up-regulated genes in *Cladocopium* sp.

Domain Name	Summary from Pfam Database	Gene Number
AAA_5	AAA domain (dynein-related subfamily)	6
DHC_N2	Dynein heavy chain, *N*-terminal region 2	5
AAA	ATPase family associated with various cellular activities	4
AAA_6	Hydrolytic ATP binding site of dynein motor region	4
TIG	IPT/TIG domain	4

Four upregulated genes in Cladocopium (Table 1) encoded proteins with TIG domains that have an immunoglobulin-like fold and are found in cell surface receptors that control cell dissociation [40,41]. This might contribute to adhesion between neighboring cells and to the extracellular matrix composition and explain some of the changes observed in cell aggregations.

There were more downregulated genes than upregulated genes in both *Symbiodinium* and *Cladocopium* (Figure 4). PPR (pentatricopeptide repeat) protein (Table 2) is involved in RNA editing [42] and extensive RNA editing has been reported in organelles of Symbiodiniaceae [43,44]. Five genes for photosynthesis were downregulated (Figure 4). These changes may explain observed reductions in photosystem efficiency in *C. goreaui* [13].

Table 2. Domains encoded by more than three down-regulated genes in *Cladocopium* sp.

Domain Name	Summary from Pfam Database	Gene Number
Ank	Ankyrin repeat	10
Ank_2	Ankyrin repeats (3 copies)	10
Ank_3	Ankyrin repeat	10
Ank_4	Ankyrin repeats (many copies)	10
Ank_5	Ankyrin repeats (many copies)	10
PPR_2	PPR repeat family	6
RCC1_2	Regulator of chromosome condensation (RCC1) repeat	6
ANAPC3 (Apc3)	Anaphase-promoting complex, cyclosome, subunit 3	5
Pkinase	Protein kinase domain	5
PPR	PPR repeat	5
PPR_3	Pentatricopeptide repeat domain	5
Abhydrolase_5	Alpha/beta hydrolase family	4
Abhydrolase_6	Alpha/beta hydrolase family	4
Lipase_3	Lipase (class 3)	4
PPR_1	PPR repeat	4
TPR_14	Tetratricopeptide repeat	4
YukD	WXG100 protein secretion system (Wss), protein YukD	4

Other downregulated gene groups were related to intracellular degradation processes, including hydrolase and lipase, and to subunit 3 of the anaphase-promoting complex/cyclosome [45]. The downregulated gene (s3282_g2) with abhydrolase and chlorophyllase domains is likely related to chlorophyll degradation [46]. The gene, s576_g21, for cell division control (CDC) protein 2 is downregulated in *Cladocopium*. Downregulation of six genes with RCC1 (regulator of chromosome condensation) and three genes with CDC domains suggest some effect on cell division. Thus, several negative consequences of nanoPS$_{42}$ exposure are suggested by DEGs (summarized in Figure 5).

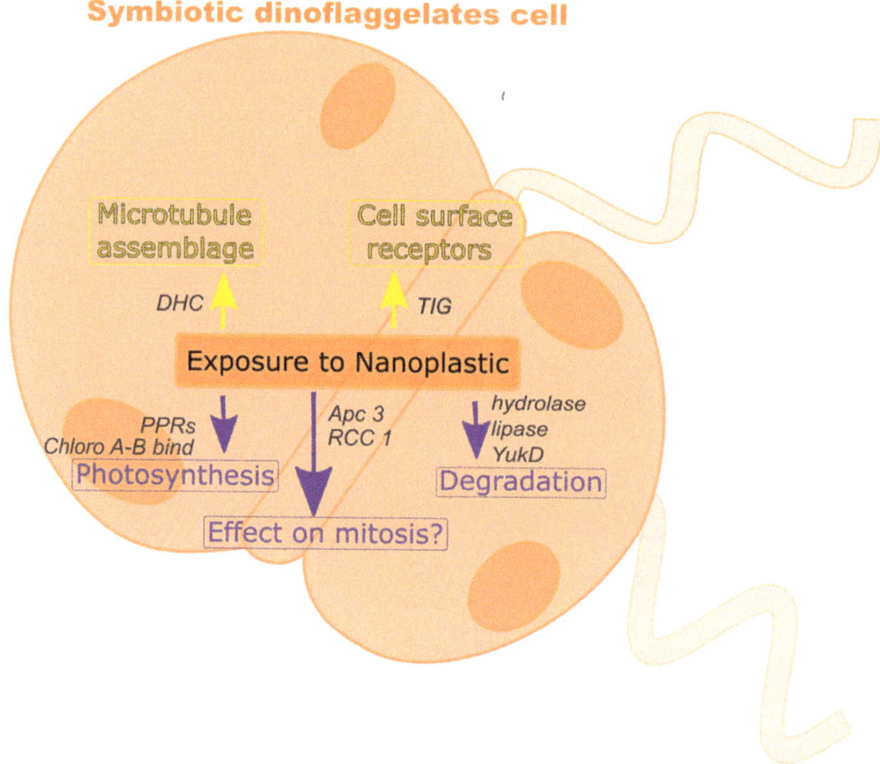

Figure 5. Exposure to nanoPS$_{42}$ changes gene expression levels in symbiotic dinoflagellates. Yellow and purple arrows show up-regulation and down-regulation of gene expression, respectively.

4. Conclusions

Previous studies have shown that nano-plastics have adverse effects on different algae groups [32,34,35,47,48], and a recent study shows that micro-plastics have similarly negative effects on an endosymbiotic dinoflagellate *Cladocopium goreaui* [13]. No previous studies have been conducted on nanoPS$_{42}$ effects on Symbiodiniaceae. We found significant changes in aggregation, photosystem efficiency, and aggregate sinking velocity of symbiotic dinoflagellates, which is coupled with variations in gene expression patterns after exposure to nanoPS$_{42}$. The reduction in photosystem efficiency and photosystem gene expression patterns could have led to the observed reduced growth rates and are especially problematic given the obligate photosymbiotic nature of the host animals of the dinoflagellates.

Supplementary Materials: The following are available online at http://www.mdpi.com/2076-2607/8/11/1759/s1. Figure S1: Cell abundance in treatment tanks, control tanks, and outside controls. Figure S2: NanoPS exposure changes aggregation behaviour, reduces cell numbers, and alters size class distributions. Table S1: Relationship between nanoPS$_{42}$ concentration and particles per tank. Table S2: Sampling days of each tank. Table S3: Genes that responded to nano-plastic exposure in *Symbiodinium tridacnidorum*. Table S4: Genes that responded to nano-plastic exposure in *Cladocopium* sp. Table S5: Differentially expressed genes with a two-fold difference between the controls and nano-plastic exposure (*Symbiodinium tridacnidorum* cladeA, TMM FPKM values, $p < 0.001$). Table S6: Differentially expressed genes with a two-fold difference between the controls and nano-plastic exposure (*Cladocopium sp.* cladeC, TMM FPKM values, $p < 0.001$).

Author Contributions: C.R. designed the study and performed the experiment. E.S. carried out RNA analyses. K.K. and C.R. performed RNA-seq mapping and cluster analyses. All authors wrote the manuscript. All authors have read and agreed to the published version of the manuscript.

Funding: This research received no external funding. This research was funded by Okinawa Institute of Science and Technology Graduate University support for the Light-Matter Interactions for Quantum Technologies (SNC) and the Marine Genomics Unit (NS).

Acknowledgments: We are grateful to the DNA sequencing section of OIST for RNA preparation and sequencing and to the OIST imaging section for 3D image support. We thank members of the MGU, especially Haruhi Narisoko for cell culturing and Kanako Hisata for IT support. We are grateful for the help and support provided by K. Deasy from the Engineering Support Section of Research Support Division at OIST. We gratefully acknowledge Síle Nic Chormaic and Noriyuki Satoh for their continuing personal and material support and kind encouragement throughout the project.

Conflicts of Interest: The authors declare no conflict of interest.

Availability of Data: Data are available in the electronic supplementary material. Raw sequence data are available from PRJNA627564 in NCBI database. Symbiodinium (currently the family Symbiodiniaceae) clade A3 and C genomes: clade A3 (https://marinegenomics.oist.jp/symb/viewer/info?project_id=37) and clade C (https://marinegenomics.oist.jp/symb/viewer/info?project_id=40). Transcript models of Symbiodinium clades A3 and C: https://marinegenomics.oist.jp/symb/download/syma_transcriptome_37.fasta.gz and https://marinegenomics.oist.jp/symb/download/symC_aug_40.fa.gz.

References

1. McKinney, M.L. Is Marine Biodiversity at Less Risk? Evidence and Implications. *Divers. Distrib.* **1998**, *4*, 3–8.

2. Hughes, T.P.; Anderson, K.D.; Connolly, S.R.; Heron, S.F.; Kerry, J.T.; Lough, J.M.; Baird, A.H.; Baum, J.K.; Berumen, M.L.; Bridge, T.C.; et al. Spatial and temporal patterns of mass bleaching of corals in the Anthropocene. *Science* **2018**, *359*, 80–83. [CrossRef] [PubMed]

3. Woodhead, A.J.; Hicks, C.C.; Norström, A.V.; Williams, G.J.; Graham, N.A.J. Coral reef ecosystem services in the Anthropocene. *Funct. Ecol.* **2019**, 1023–1034. [CrossRef]

4. Imhof, H.K.; Sigl, R.; Brauer, E.; Feyl, S.; Giesemann, P.; Klink, S.; Leupolz, K.; Löder, M.G.; Löschel, L.A.; Missun, J.; et al. Spatial and temporal variation of macro-, meso- and microplastic abundance on a remote coral island of the Maldives, Indian Ocean. *Mar. Pollut. Bull.* **2017**, *116*, 340–347. [CrossRef] [PubMed]

5. Gillibert, R.; Balakrishnan, G.; Deshoules, Q.; Tardivel, M.; Magazzù, A.; Donato, M.G.; Maragò, O.M.; De La Chapelle, M.L.; Colas, F.J.; Lagarde, F.; et al. Raman Tweezers for Small Microplastics and Nanoplastics Identification in Seawater. *Environ. Sci. Technol.* **2019**, *53*, 9003–9013. [CrossRef]

6. Tang, J.; Ni, X.; Zhou, Z.; Wang, L.; Lin, S. Acute microplastic exposure raises stress response and suppresses detoxification and immune capacities in the scleractinian coral Pocillopora damicornis. *Environ. Pollut.* **2018**, *243*, 66–74. [CrossRef] [PubMed]

7. Hall, N.M.; Berry, K.L.E.; Rintoul, L.; Hoogenboom, M.O. Microplastic ingestion by scleractinian corals. *Mar. Biol.* **2015**, *162*, 725–732. [CrossRef]

8. Connors, E.J. Distribution and biological implications of plastic pollution on the fringing reef of Mo'orea, French Polynesia. *PeerJ* **2017**, *5*, e3733. [CrossRef]

9. Axworthy, J.B.; Padilla-Gamiño, J.L. Microplastics ingestion and heterotrophy in thermally stressed corals. *Sci. Rep.* **2019**, *9*, 1–8. [CrossRef]

10. Okubo, N.; Takahashi, S.; Nakano, Y. Microplastics disturb the anthozoan-algae symbiotic relationship. *Mar. Pollut. Bull.* **2018**, *135*, 83–89. [CrossRef]

11. Hankins, C.; Duffy, A.; Drisco, K. Scleractinian coral microplastic ingestion: Potential calcification effects, size limits, and retention. *Mar. Pollut. Bull.* **2018**, *135*, 587–593. [CrossRef]

12. Reichert, J.; Schellenberg, J.; Schubert, P.; Wilke, T. Responses of reef building corals to microplastic exposure. *Environ. Pollut.* **2018**, *237*, 955–960. [CrossRef]

13. Su, Y.; Zhang, K.; Zhou, Z.; Wang, J.; Yang, X.; Tang, J.; Li, H.; Lin, S. Microplastic exposure represses the growth of endosymbiotic dinoflagellate Cladocopium goreaui in culture through affecting its apoptosis and metabolism. *Chemosphere* **2020**, *244*, 125485. [CrossRef] [PubMed]

14. Ter Halle, A.; Jeanneau, L.; Martignac, M.; Jardé, E.; Pedrono, B.; Brach, L.; Gigault, J. Nanoplastic in the North Atlantic Subtropical Gyre. *Environ. Sci. Technol.* **2017**, *51*, 13689–13697. [CrossRef]

15. Nanoplastic should be better understood. *Nat. Nanotechnol.* **2019**, *14*, 299. [CrossRef]

16. Gago, J.; Carretero, O.; Filgueiras, A.; Viñas, L. Synthetic microfibers in the marine environment: A review on their occurrence in seawater and sediments. *Mar. Pollut. Bull.* **2018**, *127*, 365–376. [CrossRef]

17. Koelmans, A.A.; Besseling, E.; Foekema, E.; Kooi, M.; Mintenig, S.; Ossendorp, B.C.; Redondo-Hasselerharm, P.E.; Verschoor, A.; Van Wezel, A.P.; Scheffer, M. Risks of Plastic Debris: Unravelling Fact, Opinion, Perception, and Belief. *Environ. Sci. Technol.* **2017**, *51*, 11513–11519. [CrossRef]

18. Galloway, T.S.; Cole, M.; Lewis, C. Interactions of microplastic debris throughout the marine ecosystem. *Nat. Ecol. Evol.* **2017**, *1*, 116. [CrossRef]

19. Mies, M.; Braga, F.; Scozzafave, M.S.; De Lemos, D.E.L.; Sumida, P.Y.G. Early development, survival and growth rates of the giant clam Tridacna crocea (Bivalvia: Tridacnidae). *Braz. J. Oceanogr.* **2012**, *60*, 127–133. [CrossRef]

20. Hirose, M.; Reimer, J.D.; Hidaka, M.; Suda, S. Phylogenetic analyses of potentially free-living Symbiodinium spp. isolated from coral reef sand in Okinawa, Japan. *Mar. Biol.* **2008**, *155*, 105–112. [CrossRef]

21. Yamashita, H.; Koike, K. Genetic identity of free-living Symbiodinium obtained over a broad latitudinal range in the Japanese coast: Phylogeny of free-living Symbiodinium. *Phycol. Res.* **2013**, *61*, 68–80. [CrossRef]

22. Schoenberg, D.A.; Trench, R.K. Genetic variation in Symbiodinium (=Gymnodinium) microadriaticum Freudenthal, and specificity in its symbiosis with marine invertebrates. I. Isoenzyme and soluble protein patterns of axenic cultures of Symbiodinium microadriaticum. In Proceedings of the Royal Society of London. Series B. Biological Sciences; The Royal Society: London, UK, 1980; 207, pp. 405–427.

23. Long, M.; Moriceau, B.; Gallinari, M.; Lambert, C.; Huvet, A.; Raffray, J.; Soudant, P. Interactions between microplastics and phytoplankton aggregates: Impact on their respective fates. *Mar. Chem.* **2015**, *175*, 39–46. [CrossRef]

24. Shanks, A.L.; Edmondson, E.W. Laboratory-made artificial marine snow: A biological model of the real thing. *Mar. Biol.* **1989**, *101*, 463–470. [CrossRef]

25. Shoguchi, E.; Beedessee, G.; Tada, I.; Hisata, K.; Kawashima, T.; Takeuchi, T.; Arakaki, N.; Fujie, M.; Koyanagi, R.; Roy, M.C.; et al. Two divergent Symbiodinium genomes reveal conservation of a gene cluster for sunscreen biosynthesis and recently lost genes. *BMC Genom.* **2018**, *19*, 1–11. [CrossRef]

26. Beedessee, G.; Hisata, K.; Roy, M.C.; Van Dolah, F.M.; Satoh, N.; Shoguchi, E. Diversified secondary metabolite biosynthesis gene repertoire revealed in symbiotic dinoflagellates. *Sci. Rep.* **2019**, *9*, 1–12. [CrossRef]

27. Arossa, S.; Martin, C.; Rossbach, S.; Duarte, C.M. Microplastic removal by Red Sea giant clam (Tridacna maxima). *Environ. Pollut.* **2019**, *252*, 1257–1266. [CrossRef]

28. Cole, M.; Lindeque, P.; Halsband, C.; Galloway, T.S. Microplastics as contaminants in the marine environment: A review. *Mar. Pollut. Bull.* **2011**, *62*, 2588–2597. [CrossRef]

29. Passow, U.; Sweet, J.; Francis, S.; Xu, C.; Dissanayake, A.; Lin, Y.; Santschi, P.; Quigg, A. Incorporation of oil into diatom aggregates. *Mar. Ecol. Prog. Ser.* **2019**, *612*, 65–86. [CrossRef]

30. Li, B.; Dewey, C.N. RSEM: Accurate transcript quantification from RNA-Seq data with or without a reference genome. *BMC Bioinform.* **2011**, *12*, 323. [CrossRef] [PubMed]

31. Robinson, M.D.; Oshlack, A. A scaling normalization method for differential expression analysis of RNA-seq data. *Genome Biol.* **2010**, *11*, R25. [CrossRef]

32. Prata, J.C.; Lavorante, B.R.; Montenegro, M.D.C.B.; Guilhermino, L. Influence of microplastics on the toxicity of the pharmaceuticals procainamide and doxycycline on the marine microalgae Tetraselmis chuii. *Aquat. Toxicol.* **2018**, *197*, 143–152. [CrossRef] [PubMed]

33. Lagarde, F.; Olivier, O.; Zanella, M.; Daniel, P.; Hiard, S.; Caruso, A. Microplastic interactions with freshwater microalgae: Hetero-aggregation and changes in plastic density appear strongly dependent on polymer type. *Environ. Pollut.* **2016**, *215*, 331–339. [CrossRef]

34. Zhang, C.; Chen, X.; Wang, J.; Tan, L. Toxic effects of microplastic on marine microalgae Skeletonema costatum: Interactions between microplastic and algae. *Environ. Pollut.* **2017**, *220*, 1282–1288. [CrossRef]

35. Nolte, T.M.; Hartmann, N.B.; Kleijn, J.M.; Garnæs, J.; Van De Meent, D.; Hendriks, A.J.; Baun, A. The toxicity of plastic nanoparticles to green algae as influenced by surface modification, medium hardness and cellular adsorption. *Aquat. Toxicol.* **2017**, *183*, 11–20. [CrossRef]

36. Logan, B.E.; Alldredge, A.L. Potential for increased nutrient uptake by flocculating diatoms. *Mar. Biol.* **1989**, *101*, 443–450. [CrossRef]

37. Andersen, K.P.; Dam, H.G. Coagulation efficiency and aggregate formation in marine phytoplankton. *Mar. Biol.* **1990**, *107*, 235–245. [CrossRef]

38. Ewilson, C.; González-Billault, C. Regulation of cytoskeletal dynamics by redox signaling and oxidative stress: Implications for neuronal development and trafficking. *Front. Cell. Neurosci.* **2015**, *9*, 381. [CrossRef]

39. Magni, S.; Della Torre, C.; Garrone, G.; D'Amato, A.; Parenti, C.; Binelli, A. First evidence of protein modulation by polystyrene microplastics in a freshwater biological model. *Environ. Pollut.* **2019**, *250*, 407–415. [CrossRef] [PubMed]

40. Collesi, C.; Santoro, M.M.; Gaudino, G.; Comoglio, P.M. A splicing variant of the RON transcript induces constitutive tyrosine kinase activity and an invasive phenotype. *Mol. Cell. Biol.* **1996**, *16*, 5518–5526. [CrossRef]

41. Bork, P.; Doerks, T.; Springer, T.A.; Snel, B. Domains in plexins: Links to integrins and transcription factors. *Trends Biochem. Sci.* **1999**, *24*, 261–263. [CrossRef]

42. Barkan, A.; Walker, M.; Nolasco, M.; Johnson, D. A nuclear mutation in maize blocks the processing and translation of several chloroplast mRNAs and provides evidence for the differential translation of alternative mRNA forms. *EMBO J.* **1994**, *13*, 3170–3181. [CrossRef]

43. Mungpakdee, S.; Shinzato, C.; Takeuchi, T.; Kawashima, T.; Koyanagi, R.; Hisata, K.; Tanaka, M.; Goto, H.; Fujie, M.; Lin, S.; et al. Massive Gene Transfer and Extensive RNA Editing of a Symbiotic Dinoflagellate Plastid Genome. *Genome Biol. Evol.* **2014**, *6*, 1408–1422. [CrossRef]

44. Shoguchi, E.; Shinzato, C.; Hisata, K.; Satoh, N.; Mungpakdee, S. The Large Mitochondrial Genome of Symbiodinium minutumReveals Conserved Noncoding Sequences between Dinoflagellates and Apicomplexans. *Genome Biol. Evol.* **2015**, *7*, 2237–2244. [CrossRef] [PubMed]

45. Peters, J.-M. The anaphase promoting complex/cyclosome: A machine designed to destroy. *Nat. Rev. Mol. Cell Biol.* **2006**, *7*, 644–656. [CrossRef]

46. Tsuchiya, T.; Ohta, H.; Okawa, K.; Iwamatsu, A.; Shimada, H.; Masuda, T.; Takamiya, K.-I. Cloning of chlorophyllase, the key enzyme in chlorophyll degradation: Finding of a lipase motif and the induction by methyl jasmonate. *Proc. Natl. Acad. Sci. USA* **1999**, *96*, 15362–15367. [CrossRef]

47. Besseling, E.; Wang, B.; Lürling, M.; Koelmans, A.A. Nanoplastic Affects Growth of S. obliquus and Reproduction of D. magna. *Environ. Sci. Technol.* **2014**, *48*, 12336–12343. [CrossRef]

48. Bellingeri, A.; Casabianca, S.; Capellacci, S.; Faleri, C.; Paccagnini, E.; Lupetti, P.; Koelmans, A.A.; Penna, A.; Corsi, I. Impact of polystyrene nanoparticles on marine diatom Skeletonema marinoi chain assemblages and consequences on their ecological role in marine ecosystems. *Environ. Pollut.* **2020**, *262*, 114268. [CrossRef]

Publisher's Note: MDPI stays neutral with regard to jurisdictional claims in published maps and institutional affiliations.

 *microorganisms*

MDPI

Article

Assessing the Use of Molecular Barcoding and qPCR for Investigating the Ecology of *Prorocentrum minimum* (Dinophyceae), a Harmful Algal Species

Kate McLennan, Rendy Ruvindy, Martin Ostrowski and Shauna Murray *

Faculty of Science, University of Technology Sydney, Ultimo, NSW 2007, Australia;
kate.mclennan@alumni.uts.edu.au (K.M.); rendy.ruvindy@uts.edu.au (R.R.); martin.ostrowski@uts.edu.au (M.O.)
* Correspondence: Shauna.Murray@uts.edu.au

Abstract: *Prorocentrum minimum* is a species of marine dinoflagellate that occurs worldwide and can be responsible for harmful algal blooms (HABs). Some studies have reported it to produce tetrodotoxin; however, results have been inconsistent. qPCR and molecular barcoding (amplicon sequencing) using high-throughput sequencing have been increasingly applied to quantify HAB species for ecological analyses and monitoring. Here, we isolated a strain of *P. minimum* from eastern Australian waters, where it commonly occurs, and developed and validated a qPCR assay for this species based on a region of ITS rRNA in relation to abundance estimates from the cultured strain as determined using light microscopy. We used this tool to quantify and examine ecological drivers of *P. minimum* in Botany Bay, an estuary in southeast Australia, for over ~14 months in 2016–2017. We compared abundance estimates using qPCR with those obtained using molecular barcoding based on an 18S rRNA amplicon. There was a significant correlation between the abundance estimates from amplicon sequencing and qPCR, but the estimates from light microscopy were not significantly correlated, likely due to the counting method applied. Using amplicon sequencing, ~600 unique actual sequence variants (ASVs) were found, much larger than the known phytoplankton diversity from this region. *P. minimum* abundance in Botany Bay was found to be significantly associated with lower salinities and higher dissolved CO_2 levels.

Keywords: *Prorocentrum minimum*; harmful algae; next-generation sequencing

Citation: McLennan, K.; Ruvindy, R.; Ostrowski, M.; Murray, S. Assessing the Use of Molecular Barcoding and qPCR for Investigating the Ecology of *Prorocentrum minimum* (Dinophyceae), a Harmful Algal Species. *Microorganisms* **2021**, *9*, 510. https://doi.org/10.3390/microorganisms9030510

Academic Editor: Thomas Mock

Received: 20 January 2021
Accepted: 19 February 2021
Published: 28 February 2021
Corrected: 26 September 2022

Publisher's Note: MDPI stays neutral with regard to jurisdictional claims in published maps and institutional affiliations.

1. Introduction

In recent decades, there has been an apparent global increase in the range, intensity, and frequency of harmful algal blooms (HABs) linked to a variety of factors, including range expansions, increases in anthropogenic nutrients into coastal water bodies, and increased aquaculture [1–6]. *Prorocentrum minimum* is a planktonic marine dinoflagellate that forms HABs and is found commonly in temperate estuarine and coastal waters [7]. *P. minimum* blooms are most common in eutrophic coastal waters of the northern hemisphere; however, they have also been reported in tropical and subtropical regions globally [1,7–10]. Although few studies have been conducted on *P. minimum* in Australia, it is known to occur in high abundances in some regions, with frequent blooms in the Hawkesbury River in New South Wales (NSW) [11]. In line with the global increase of HABs, *P. minimum* appears to have expanded its geographical range over the last 40 years [1,10,12]. *P. minimum* usually blooms in warm brackish waters that are heavily impacted by excess nutrients, which has led to its presence being used as an indicator of eutrophication in water bodies in the northern hemisphere [1,11,13].

The abundance and even dominance of *P. minimum* in dynamic estuarine and coastal systems may be due to its broad salinity tolerance range of 5–17 PSU [9,14] and broad temperature tolerance range of 3–30 °C [1,15,16]. *P. minimum* typically blooms in low-turbulence environments during periods of high irradiance levels [1]; however, it has been demonstrated that the species can survive complete darkness for extended periods [17], which may allow it to survive in ship ballast waters. *P. minimum* is considered to be a mixotroph, able to supplement its nutrient intake due to feeding on smaller microbes, such as *Cryptomonas* spp., in response to depleted nutrients in the water [7,11,18]. Despite the ability to survive with low nutrients, *P. minimum* preferentially grows in water bodies with high nutrient loadings, typical of eutrophic water bodies [1,9]. *P. minimum* growth has been found to be associated with high inorganic nitrogen (N) and phosphorus (P), strongly linked to anthropogenic sources, such as fertilisers [7].

P. minimum blooms have been associated with several different marine biotoxins [19–22]; however, the identities of the compounds and their modes of toxicity are debated. *P. minimum* blooms have shown toxic effects on shellfish, including mortality, poor development, and altered behaviours [1,23–25]. Recently, a *P. minimum* bloom has been associated with the neurotoxin tetrodotoxin (TTX) [19,26,27], possibly due to bacterial species associated with *P. minimum* [26,28]. It has been suggested that *P. minimum* toxicity is variable depending on the strain of the species and the environmental conditions under which it is grown [1,26]. Due to incidences of toxin accumulation in shellfish and the impacts on shellfish growth of *P. minimum* toxins, it is an important HAB species to monitor in shellfish-harvesting regions [1,11,23,24,26].

Until recently, light microscopy has been the only routine method available to identify and manually count HAB species [29–32]. However, this method is relatively time-consuming, requires a very high level of taxonomic expertise, and is not able to identify cryptic species, which may appear morphologically indistinguishable from one another despite toxicological differences. For these reasons, alternative methods of monitoring HABs have been developed. Molecular genetic techniques can provide rapid and sensitive HAB monitoring [29,30,33,34]. Two molecular genetic methods used are quantitative polymerase chain reaction (qPCR) and molecular barcoding. Molecular barcoding, also referred to as amplicon sequencing, is becoming invaluable in studying marine ecological assemblages, as it allows for uncultured cells in samples to be identified [29,35,36]. However, due to the existence of variability in the copy numbers of genes among microalgal species, particularly in dinoflagellates [37–39], the number of gene copies amplified may not reflect the relative abundance of species in the sample. There is also a bias introduced with the use of broad-range primers, which can lead to certain sequences being preferentially amplified, giving a skewed proportional abundance of target species [29,40]. For this reason, the quantification of dinoflagellate species using amplicon sequencing is uncertain and not accurate when compared with other methods, including qPCR and light microscopy [29,41]. The use of molecular barcoding, which provides an overview of the genetic composition of microbial communities, in conjunction with qPCR, may improve the quantitative assessment of the impact of HAB species in the context of, for example, seasonal changes in the wider microbial community.

The aim of the study was to develop and assess new molecular genetic approaches to investigate the dynamics, community, and environmental drivers of *P. minimum* in an Australian estuary. To do this, a local isolate of *P. minimum* from Australia was established, and qPCR approaches were designed and tested for the detection and quantification of *P. minimum*. In addition, 18S rRNA amplicon sequences from estuarine water samples were examined to compare the specificity, detection limits, and quantification accuracy of the methods. Environmental samples, including physico-chemical data, were collected monthly for 14 months from 2016 to 2017 from two sites in Botany Bay, an estuary in southeast Australia. Data of the entire microbial community, the abundance of *P. minimum*, and the corresponding physico-chemical variables were examined to assess the factors impacting the presence and abundance of *P. minimum* in an Australian estuary.

2. Materials and Methods

2.1. Sampling Sites

Fortnightly, phytoplankton samples were collected from two sites, Towra Point (−34.007, 151.19) and Bare Island (−33.992, 151.23), which are both located in Botany Bay, a heavily modified estuary in southeast Australia (Figure 1), as part of the coastal and benthic sampling for the Marine Microbes project, conducted by Bioplatforms Australia (BPA) [42]. Water samples were also collected from Towra Point as part of the NSW Food Authority's Shellfish Safety program for the purpose of identification and enumeration of phytoplankton. A phytoplankton net was towed to collect a dense sample to verify species identity by microscopy. Lugol's iodine (elemental iodine (5%) and potassium iodide (KI, 10%) and distilled water used at 1 mL/50 mL sample) was added immediately after collection to preserve cells [43]. In the laboratory, gravity-assisted membrane filtration was used to concentrate samples, and cell counts were completed using a Sedgewick Rafter counting chamber following a previously published protocol [43–45]. Highly toxic species were counted to a minimum level of detection of 50 cells^{-L}, while others, including *P. minimum*, were counted to a minimum level of detection of 500 cells^{-L}. All counts were completed using Zeiss Axiolab or Zeiss Standard microscopes with a maximum magnification of 1000×. All cells were identified to the nearest taxon able to be accurately identified, and if separation of species was not possible, the cells were assigned to a species group.

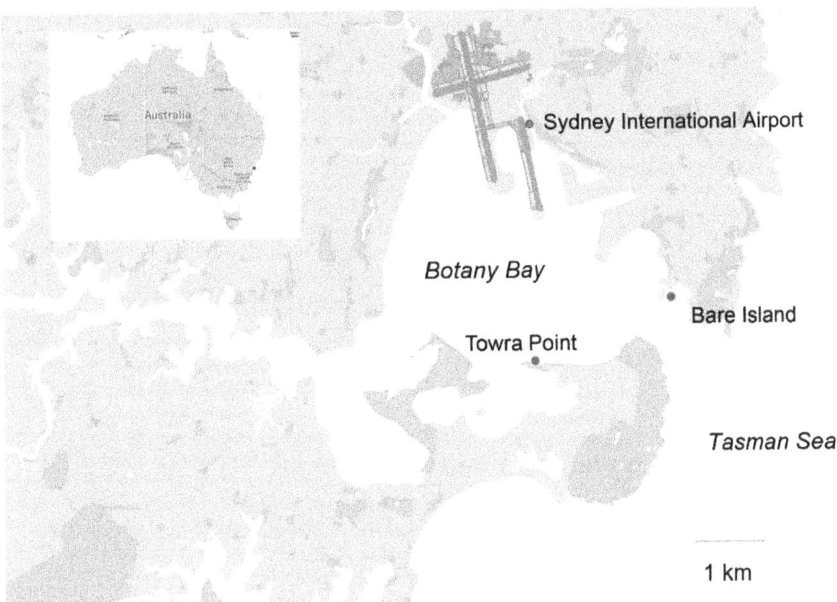

Figure 1. Map of Botany Bay in southeast Australia, highlighting the two sampling points used for the Marine Microbes project, Bare Island and Towra Point.

2.2. Sampling, DNA Extraction for Amplicon Sequencing, Metabarcoding, and Physico-Chemical Data

All samples followed standard operating procedures (SOPs) outlined in Sections 1.5.1 and 1.7 of the Australian Microbiome Scientific Manual (version 2020) [46]. Briefly, 2 L samples were filtered in triplicate through 0.22 μm dia. pore-size polyethersulfone filters (Sterivex™, Merck KGaA, Darmstadt, Germany) to concentrate algal and bacterial cells. DNA was extracted using the QIAGEN DNeasy PowerSoil Kit according to the manufacturer's instructions (QIAGEN, Hilden, Germany). Amplicon sequencing was completed using the HTS Illumina MiSeq technology at the Ramaciotti Centre for Genomics, University of New South Wales (UNSW). Samples were collected fortnightly at Towra Point and monthly at Bare Island. A series of $4\times$ samples were collected during and after a rainfall event at both sites from 3 to 14 June 2016.

2.3. Cell Culture and Culturing Conditions

Several 600 mL water samples were collected from Berowra Creek (-33.5342, 151.1459), a tributary of the Hawkesbury River in NSW, about 50 km north of Botany Bay, on 12 April 2019 following reports of a bloom of this species on 8 April 2019. The bloom was reported by Ana Rubio from Hornsby Shire Council, who regularly monitors and samples Berowra Creek. Clonal cultures of cells with the morphological features typical of *P. minimum* were isolated under a light microscope using a micropipette and cultured in a 96-well plate format. The Berowra Creek sample was successful for the isolation of *P. minimum*, and one isolate survived and was successively inoculated into 50 mL of media. The clonal isolate was grown in K media [47] in an incubator at 18 °C at a salinity of 35 PSU under a 12/12 h light cycle (±100 µmol photon m^{-2} s^{-1}). A salinity of 35 PSU closely matches the salinity found at Berowra Creek where the strain was isolated from. A cultured strain of a closely related species, *Prorocentrum* cf. *balticum* (obtained from the Cawthron Institute Culture Collection, Nelson, New Zealand, CICCM, CAWD38), was used as a negative control in the qPCR assay. *P.* cf. *balticum* was maintained in culture in identical conditions to *P. minimum* in K media [47]. *Prorocentrum lima* (SM43), *Prorocentrum concavum* (SM46), and *Prorocentrum cassubicum* (CS881, from the Australian National Algal Culture Collection) cultures were grown in F/10 media [48] under a 12/12 h light cycle. *P. concavum* and *P. lima* were incubated at 28 °C, and *P. cassubicum* at 25 °C.

2.4. Toxin Analysis

To prepare samples for toxin analysis using liquid chromatography–mass spectrometry (LC–MS, ThermoFisher Scientific Q Exactive, Waltham, MA, USA), 20 mL of a dense (28,000 cells mL^{-1}) culture was centrifuged at $4000\times g$ for 10 min to form a pellet, and the supernatant was discarded. The sample was then freeze-dried and stored at -20 °C. Analysis of TTX presence in the culture was completed by Dr. Chowdhury Sarowar at the Sydney Institute of Marine Science (SIMS) following an adapted method from [49].

Briefly, 5 mL of 1 mM acetic acid was added to the sample and vortexed for 90 s, after which the samples were placed in a boiling water bath for 5 min and then cooled to room temperature. The cooled sample was placed in an ultrasonic bath for 1 min and then centrifuged to pellet the cellular debris, and the supernatant was used with or without dilution for LC–MS analysis. A Thermo Scientific™ Q EXACTIVE™ MS (Waltham, MA, USA) was used for the detection of TTX. The source parameters for detection were as follows: sheath gas and auxiliary gas flow rates of 50 and 13, respectively (arbitrary units); a spray voltage of 3.5 kV; a capillary temperature of 263 °C; and an auxiliary gas heater temperature of 425 °C.

Chromatographic separation was performed on a Thermo Scientific™ ACCELA™ UPLC system (Waltham, MA, USA). Analysis was performed using an Acquity UPLC BEH Phenyl 1.7 µm 100 × 2.1 mm column with an injection volume of 5 µL. The mobile phases used were A (water/formic acid/NH4OH at 500:0.075:0.3 v/v/v) and B (acetonitrile/water/formic acid at 700:300:0.1 v/v/v). Initial condition starts at A/B 2:98 at a flow rate of 400 µL/min and held for 5 min. The condition was then linearly changed for over 3.5 min from A/B (2:98) to A/B (50:50). The flow rate was then changed from 400 µL/min to 500 µL/min for over 2 min. The chromatographic condition was then rapidly changed to initial buffer conditions A/B (2:98) for over 0.5 min, while the flow rate was kept at 500 µL/min. The flow rate was then increased to 800 µL/min for over 0.5 min and held for 0.6 min. The flow rate was then decreased back to the initial flow rate of 400 µl/min, and the condition changed to A/B 100:0. A certified standard solution of TTX was sourced from Enzo Life Sciences (Exeter, UK).

2.5. DNA Extraction and PCR for Strain Identification

Cell density was determined by enumeration with Lugol's iodine-stained cells using a Sedgewick Rafter counting chamber [44,45]. Following microscopic counts, samples of the *Prorocentrum* spp. cultures were harvested by centrifugation at $4000\times g$ for 10 min to be used for DNA extraction. DNA was extracted from the *P. minimum* (and other *Prorocentrum* spp. to be used for negative controls) cell pellets using the QIAGEN DNeasy PowerSoil Kit (QIAGEN, Hilden, Germany) according to the manufacturer's protocol. The samples were eluted in 100 µL of buffer and stored at -20 °C until analysis. The quantity and quality of the extracted DNA were measured with a NanoDrop spectrophotometer (ThermoFisher Scientific, Waltham, MA, USA). Following DNA extraction, PCR amplification was completed using appropriate primers for the LSU rRNA and ITS rRNA regions. PCR amplification was conducted using the Bio-Rad T100 Thermal Cycler [50] (Bio-Rad Laboratories Inc., Hercules, CA, USA), targeting two rRNA regions. The LSU rRNA region was run using d1F (F) and d3B (R) primers, and the ITS rRNA region was run with ITSfwd (F) and ITSrev (R) primers (Table 1). PCR was run in 25 µL reactions with 12.5 µL of ImmoMix (Bioline, Sydney, NSW, Australia), 1 µL of BSA (bovine serum albumin), 7.5 µL of sterile water, 1.5 µL of forward primer (10 µM), and 1.5 µL of reverse primer (10 µM). The protocol used for PCR was 10 min at 95 °C, followed by 35 cycles of 95 °C for 20 s, 57 °C for 30 s, and 72 °C for 1 min, then held at 72 °C for 7 min. DNA fragments were cleaned using the DNA Clean and Concentrator (Zymo Research, Irvine, CA, USA) according to the manufacturer's protocol and sequenced at Macrogen (Seoul, Korea). Contigs were formed using the sequences in Geneious Prime (v2019.2.1, Biomatters, Ltd., Auckland, NZ). Following the assembly of the contigs for each gene region, each sequence was uploaded to NCBI BLAST nucleotide sequence search to identify the strain and confirm that it was *P. minimum* (Tables 2 and 3).

Table 1. Names and sequences of all the primers used in this project. Primers 200F and 525R are taken from [51]. The other primers were designed using the online Primer-BLAST software (NCBI). All reverse primers are in reverse complement.

Name	Sequence (Forward)	Name	Sequence (Reverse)
Primer Sequences for qPCR			
Pm 200F	TGTGTTTATTAGTTACAGAACCAGC	Pm 525R	AATTCTACTCATTCCAATTACAAGACAAT
1F Pmin	CGCAGCGAAGTGTGATAAGC	1R Pmin	TCTGGAAAGGCCAGAAGCTG
2F Pmin	TCGGCTCGAACAACGATGAA	2R Pmin	AAGCGTTCTGGAAAGGCCAG
3F Pmin	TTCTGGCCTTTCCAGAACGC	3R Pmin	CATGCCCAACAACAAGGCAA
4F Pmin	CGTATACTGCGCTTTCGGGA	4R Pmin	CACACAGAAACACACAAGCGT
5F Pmin	CCTTTCCAGAACGCTTGTGTG	5R Pmin	CTGGGCACTAGACAGCAAGG
6F Pmin	CAGGCTCAGACCGTCTTCTG	6R Pmin	AGCGTTCTGGAAAGGCCAG
7F Pmin	CAACAGTTGGTGAGGCTCT	7R Pmin	ATTCAAAAACACAGAAGATCAGGAA
8F Pmin	AACAACAGTTGGTGAGGCTCTG	8R Pmin	CAAAAACACAGAAGATCAGGAAGAC
9F Pmin	GTGAGGCTCTGGGTGGG	9R Pmin	CAAAAACACAGAAGATCAGGAAGAC
10F Pmin	TCATTCGCACGCATCCATTC	10R Pmin	AAGGACAGGCACAGAAGACG
11F Pmin	TTCAGTGCACAGGGTCTTCC	11R Pmin	GTCTTGGTAGGAGTGCGCTG
12F Pmin	GCCTTTCCAGAACGCTTGTGT	12R Pmin	GCTGACCTAACTTCATGTCTTGG
13F Pmin	CGCTTGTGTGTTTCTGTGTG	13R Pmin	CCATGCCCAACAACAAGGC
14F Pmin	TCTTCCCACGCAAGCAACT	14R Pmin	CGGGTTTGCTGACCTAAACT
15F Pmin	ACATTCGCACGCATCCATTC	15R Pmin	TTGCTGCCCTTGAGTCTCTG
16F Pmin	AACAGTTGGTGAGGCTCTGG	16R Pmin	AAGGACAGGCACAGAAGACG
17F Pmin	ACAACAGTTGGTGAGGCTCT	17R Pmin	TTGCTGCCCTTGAGTCTCTG
18F Pmin	CAGTTGGTGAGGCTCTGGG	18R Pmin	CAGAAGACGGTCTGAGCCTG

Table 1. *Cont.*

Name	Sequence (Forward)	Name	Sequence (Reverse)
	Primer Sequences for qPCR		
19F Pmin	TTCAGTGCACAGGGTCTTCC	19R Pmin	CATGCCCAACAACAAGGCAA
20F Pmin	ATTCCAGCTTCTGGCCTGTC	20R Pmin	TAGTTGCTTGCGTGGGAAGA
21F Pmin	CTGTCCAGAACGCTTGTGTG	21R Pmin	CTTCTAGTTGCTTGCGTGGG
22F Pmin	TTCCCACGCAAGCAACTAGA	22R Pmin	GCACTAGACAGCAAGGCCA
	Primer Sequences for Amplicon Sequencing		
Modified TAReuk454FWD1	CCAGCASCYGCGGTAATTCC	Modified TAReukREV3	ACTTTCGTTCTTGATYRATGA
	Primer Sequences for PCR and Sanger Sequencing		
d1f	ACCCGCTGAATTTAAGCATA	d3b	TCGGAGGGAACCAGCTACTA
ITSfwd	TTCCGTAGGTGAACCTGCGG	ITSrev	ATATGCTTAAATTCAGCGGGT

Table 2. Top 5 BLAST nucleotide sequence hits of the LSU rRNA sequence of *Prorocentrum minimum* from Berowra Creek, CAWD359, as compared with sequences of *P. minimum* strains on the NCBI database, including linked accession numbers.

Strain Description	Percent Identity	Accession
Prorocentrum minimum strain DAB02 28S	99.66%	KU999985.1
Prorocentrum minimum strain D-127	99.66%	JX402086.1
Prorocentrum minimum isolate PIPV-1	99.54%	JQ616823.1
Prorocentrum minimum isolate SERC	99.54%	EU780639.1
Prorocentrum minimum strain Pmin1	99.54%	AY863004.1

Table 3. Top 5 BLAST nucleotide sequence hits of the ITS rRNA sequence of *P. minimum* from Berowra Creek, CAWD359, as compared with sequences of *P. minimum* strains on the NCBI database.

Strain Description	Percent Identity	Accession
Prorocentrum minimum strain D-127	99.67%	JX402086.1
Prorocentrum minimum strain AND3V	100.00%	EU244473.1
Prorocentrum minimum isolate PIPV-1	99.35%	JQ616823.1
Prorocentrum minimum strain PMDH01	99.35%	DQ054538.1
Prorocentrum minimum strain NMBjah049	99.67%	KY290717.1

2.6. qPCR Assay Development

2.6.1. Primer Design

A published set of primers designed for a *P. minimum*-specific qPCR assay was tested [51]. Forward (F) and reverse (R) primers were designed to amplify a 325 bp region from the partial 18S rDNA sequence of *P. minimum* accessed from GenBank (AY421791.1) (Table 1). Twenty-two new sets of primers were designed after the above primer did not pass testing using the NCBI Primer-BLAST tool, targeting ITS regions 1 and 2 of partial sequence of the *P. minimum* strain CCMP698 (EU927537.1). The sizes of the qPCR products were from 70 to 130 bp in length, with the primers 20 to 25 bp in length, with the optimal T_m (melting temperature) set at 60 °C (Table 1). All sets were expected to be specific to the target sequence of *P. minimum*, as they were compared with all available sequences in the NCBI database and were unique to *P. minimum*. All 22 primer sets were run using an identical protocol (see qPCR Assays) with *P. minimum* DNA to determine the most sensitive and efficient assay. All valid primer sets were then subjected to specificity testing with other *Prorocentrum* spp. DNA. Primer sets that amplified other *Prorocentrum* spp. were disregarded, and then standard curve testing followed.

2.6.2. qPCR Assays

qPCR was conducted using a 20 μL mix with 1 μL of DNA, 10 μL of Bio-Rad iTaq Universal SYBR Green Supermix (Bio-Rad Laboratories Inc., Hercules, CA, USA), 1 μL (at 10 μM concentration) of each of the forward and reverse primers, and 7 μL of sterile water. qPCR was performed with the following thermal cycling program of 95 °C for 10 min, followed by 40 cycles of 95 °C for 15 s and 60 °C for 1 min. A melt curve was performed on all runs, from 55 °C to 95 °C in 5 °C increments for 0.05 s each. All qPCR analyses and testing were run on the Bio-Rad CFX96 Touch Real-Time PCR Detection System in 96-well plates with a clear seal or clear plastic strips [52] (Bio-Rad Laboratories Inc., Hercules, CA, USA),). The previously published primers [51] and primer sets 1–12 followed the original qPCR protocol. Primers 13–22 were run with a modified protocol of 95 °C for 2 min, followed by 35–40 cycles of 95 °C for 15 s and 60 °C for 30 s. All assays were run using a generic thermal profile, as per above. The assays were tested for sensitivity using DNA extracted from the *P. minimum* cultures in duplicate and two negative controls containing no DNA (no template control (NTC)). Cross-reactivity of the primers was tested by running each assay on four other *Prorocentrum* spp. as negative controls. This step is crucial to ensure that the primers would only bind to *P. minimum* specifically. For other species, see Table 4. Unique primer sets developed using the NCBI Primer-BLAST tool were subjected to identical testing used for the published primer set [51]. Primer sets that were unable to amplify *P. minimum* were not sensitive (amplified past 35 cycles); those that amplified other *Prorocentrum* spp. at similar C_q values to *P. minimum* and those that had efficiencies outside 90–110% were disregarded. Those primer sets that passed specificity tests were then tested for their efficiency using a dilution series of gBlocks® gene fragments (IDT, USA) of the ITS region and *P. minimum* DNA, both with known concentrations. Standard curves were created using a 10-fold dilution series over five different concentrations and plotted with the threshold cycle (C_q) (*x*-axis) and natural log of concentration (cells/μL).

The curves were used to calculate the efficiency (E) of the assay using $E = -1 + 10^{\left(-\frac{1}{slope}\right)}$. Efficiency of qPCR assays should fall between 90% and 110% [53]. This standard curve will also be used to quantify the amount of *P. minimum* cells in unknown concentrations from environmental samples [54,55]. The assay that had acceptable efficiency and specificity was used for analysing BPA samples. After the development of the assay, qPCR was run on all extracted DNA samples from Towra Point and Bare Island collected during the BPA project from 2016 to 2017 using only one of the primer sets that passed efficiency and specificity testing. If amplification was found in the no template control (NTC), a conservative cutoff value of 3.3 C_q points or more below the C_q of the NTC was set to accept the amplification of *P. minimum* in the samples. These data on the distribution and abundance of *P. minimum* were then compared with results obtained using light microscope identification and amplicon sequencing.

Table 4. All species used as negative controls for specificity testing of the *P. minimum* qPCR assay. Species names and which culture collection they can be found are listed, as well as the strain ID.

Species Name	Culture Collection	Strain ID #
Prorocentrum cf. balticum	Cawthron Culture Collection (CICCM)	CAWD38
Prorocentrum cassubicum	Australian National Culture Collection (ANAAC)	CS-881
Prorocentrum concavum	Seafood Safety Team, University of Technology Sydney (UTS)	Pmona (SM46)
Prorocentrum lima	Seafood Safety Team, University of Technology, Sydney (UTS)	SM43

2.7. Bioinformatic Analysis

Samples collected as part of the Marine Microbes project were subjected to amplicon sequencing using primer sets to target different organisms: eukaryotes, bacteria, archaea, and fungi (Table 1). The primer sets used targeted the V4 region of 18S rRNA (Table 1) found in all eukaryotes [56]. After sequencing, the reads were trimmed, merged, concatenated, and taxonomically classified using the Earth Microbiome Project (EMP) protocol [57,58]. Following this, the resulting actual sequence variants (ASUs) were assigned to taxonomic lineages and species using the PR2 taxonomic database (version 4.12) [59] with the DADA2 (version 1.16.0) [60] assignTaxonomy and assignSpecies functions in R. The resulting data were used to extract the occurrence and relative abundance of *P. minimum* in the sequence samples; these data were used to compare with the qPCR results. The classified data were also used to discover phytoplankton species that co-occur with *P. minimum*. The relative abundance of *P. minimum* was multiplied by 100 (1 µL was used for amplicon sequencing) to give the approximate abundance in the 2 L original sample and then divided by 2 to give cells L^{-1} to be able to compare with microscope count and qPCR abundance data. Data are submitted as Supplementary Material, Table S2: Towra Point ASVs and Table S3: Bare Island ASVs.

2.8. Environmental Parameters

Physico-chemical data were collected during each sampling point according to the SOPs laid out in Australian Microbiome Scientific Manual Section 1.5. [46]. The physical parameters collected were temperature (°C), salinity (PSU), dissolved oxygen (% and mg/L), conductivity (s/m), total alkalinity (µmol/kg), and pH. The nutrients measured in the samples were nitrite, nitrate, oxidised nitrogen, phosphate, ammonium (all in µg/L), and total carbon dioxide (in µmol/kg). These data were statistically analysed with the qPCR *P. minimum* abundance data.

2.9. Statistical Analysis

To test for relationships between *P. minimum* and environmental variables, the data were first checked for normality using the Shapiro–Wilks test due to the small dataset. After failing normality ($p < 0.05$), all variables were log-transformed and tested again for normality. Several variables remained non-normally distributed ($p < 0.05$), so a nonparametric testing approach was applied. The two sites were not found to show any significant differences ($p > 0.05$), so data were pooled for both sites for further analysis. Due to the disparate nature of the dataset, multiple regression was deemed inappropriate. Instead, Kendall's tau-b correlation was used as it is more suitable for nonparametric small datasets and is more robust to error [61]. It was found to be a suitable analysis to determine preliminary relationships that could be investigated with further higher temporal data. A two-tailed correlation was run between all variables to assess the correlation between *P. minimum* abundance and environmental variables. Analyses were run in SPSS (v.26, IBM Corp, Armonk, NY, USA).

To determine whether there were any significant correlations between *P. minimum* and other phytoplankton species, co-occurrence analysis was run using the probabilistic model developed by Veech (2013), which is included in the R package "cooccur" [62,63]. This method tests all possible pairwise associations between species across samples/sites, and the output is the probability that two species co-occur at a level that is more or less frequent than the observed frequency of co-occurrence [64]. The output provides information specific to *P. minimum* and its associations, as well as the number of random and significant associations between all species in the dataset. The amplicon sequencing output was used to create a presence/absence matrix with all phytoplankton species across all the BPA sampling dates to use in the analysis.

3. Results

3.1. Strain Isolation, Identification, and Toxin Testing

A strain of *P. minimum* was successfully isolated from a water sample from Berowra Creek, NSW, in April 2019. It was initially identified as *P. minimum* based on light microscopy. The strain was also identified as *P. minimum* based on sequencing of rRNA barcoding regions, as sequencing of the LSU (GenBank accession number MT856373) and ITS rRNA regions (GenBank accession number MT895109) matched eight different *P. minimum* strains (>99.5%) when queried against the NCBI nonredundant database using blastn (Tables 3 and 4). The strain has been submitted to the Cawthron Institute Culture Collection (http://cultures.cawthron.org.nz/ (accessed on 12 February 2021)) as strain number CAWD359. The strain was tested for the presence of tetrodotoxin using a TTX standard. No TTX was detected, while TTX was detected in the spiked positive control.

3.2. qPCR Assay Development and Testing

A previously published assay with specific primers for *P. minimum* [51] was tested. This assay did not pass initial testing due to the amplification of other *Prorocentrum* species tested: *P.* cf. *balticum*, *P. lima*, *P. cassubicum*, and *P. concavum*. The assay was also not able to distinguish between products in the melt curve analysis (Table 5). The assay had a low efficiency of 70%. Following this, 22 new primer sets were tested (Table 5) to find one that was specific to *P. minimum*, sensitive, and efficient (90% < E < 110%). Only one primer set (20, Pmin 20F and Pmin 20R) displayed acceptable specificity and efficiency: E = 101% for *P. minimum* standard curve using DNA extracted from our strain (from 1.91×10^4 to 1.91×10^0 cells, Figure 2) and E = 99.3% for the standard curve using the gBlock synthetic gene fragment of the ITS region of *P. minimum* (from 1.64×10^7 to 1.64×10^3 copies, Figure 2) at an annealing temperature of 60 °C. Although this primer set was found to amplify other *Prorocentrum* species, this amplification occurred at similar or higher Cq values than that of the lowest *P. minimum* dilution point, which was the DNA equivalent of ~1.9 cells of *P. minimum*. This was even though all samples contained the DNA equivalent of $>10^4$ cells of that *Prorocentrum* species. We considered this to be an acceptable level of cross-reactivity. The assay had a reliable detection limit of ~13 cells L^{-1} when only values of at least 3.3 Cq points or less than the NTC were taken into consideration [65,66]. Primer set 20 was then used to analyse the abundance of *P. minimum* in environmental samples collected from Botany Bay.

Table 5. Specificity and efficiency testing of each primer set including the previously published one [51]. (+) and (−) mean amplification or no amplification, respectively, and (/) denotes amplification at a high C_q and/or was at or below the lowest dilution point on the *P. minimum* standard curves and/or was amplified but had a different melt peak. N/A means the primer set was not subjected to that test.

Primer Set	Specificity					Efficiency	
	P. minimum	*P. cf. balticum*	*P. cassubicum*	*P. concavum*	*P. lima*	gBlock (%)	*P. min* DNA (%)
Pm 200F/525R	+	+	−	/	/	70	−
3	+	/	−	−	−	65	−
4	+	+	−	−	−	64	−
5	+	+	+	+	−	62	−
6	+	+	+	+	+	65	−
7	+	/	−	−	−	57	−
8	+	+	−	−	−	56	−
9	+	+	−	−	−	58	−
10	+	+	N/A	/	/	60	−
11	+	−	N/A	−	−	76	−
12	+	/	N/A	/	−	85	−
13	+	−	+	N/A	+	43	−
15	+	+	N/A	+	+	328	335
19	+	+	N/A	+	+	220	383
20	+	/	−	/	/	99	101
21	+	+	+	+	+	N/A	N/A
22	+	+	+	+	+	115	147

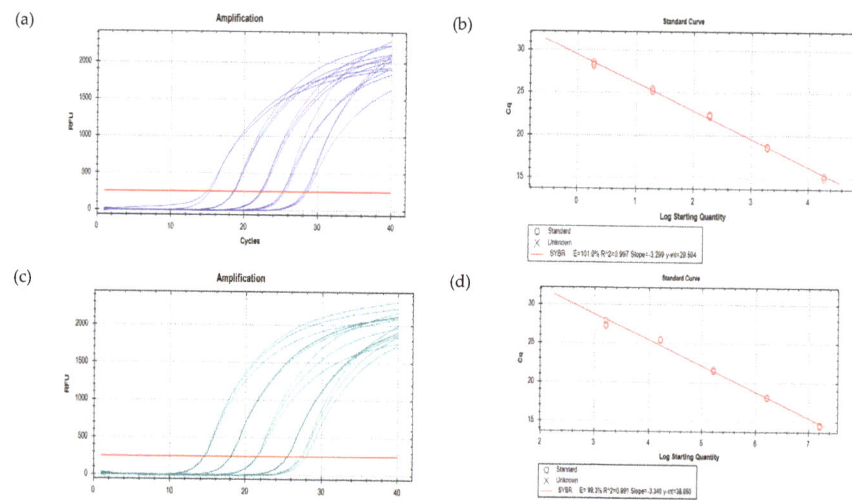

Figure 2. (**a**) Amplification plot showing dilution series using *P. minimum* culture DNA using primer set 20 and (**b**) DNA standard curve with R^2 value = 0.997 and E = 101%. (**c**) Amplification plot showing dilution series using *P. minimum* gBlocks using primer set 20 and (**d**) gBlock standard curve with R^2 value = 0.991 and E = 99.3%.

3.3. Comparison of qPCR, Light Microscope Count, and Amplicon Sequencing Abundance Results

During 2016–2017, *P. minimum* was recorded using the qPCR assay on 21 out of 27 sampling dates for Towra Point and 7 out of 17 dates for Bare Island, the two sites in Botany Bay (Figures 3 and 4). Abundances varied from 0 cells L^{-1} to 8100 cells L^{-1} at Towra Point and from 0 cells L^{-1} to 14,800 L^{-1} at Bare Island. The highest peaks were recorded in early June for Towra Point and Bare Island (Figures 3 and 4). No *P. minimum* was found on the 9 sampling dates between July 2016 and October 2016 at Bare Island (Figure 4).

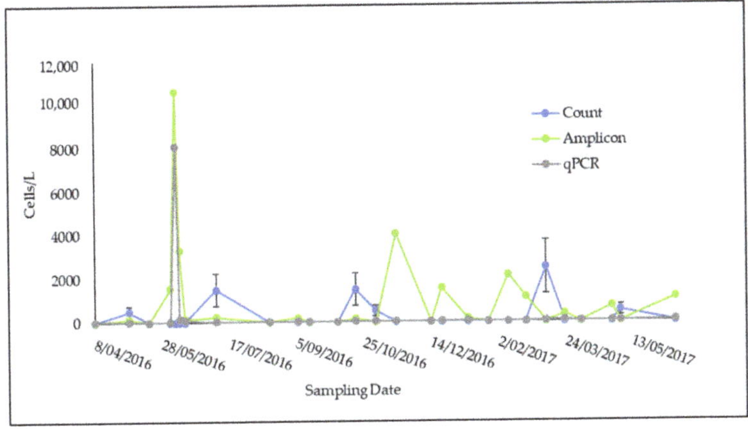

Figure 3. Towra Point *P. minimum* cell counts L^{-1} with amplicon sequencing, qPCR, and microscope counts across the BPA sampling period. Standard error bars are included for qPCR (too small to detect) and count data.

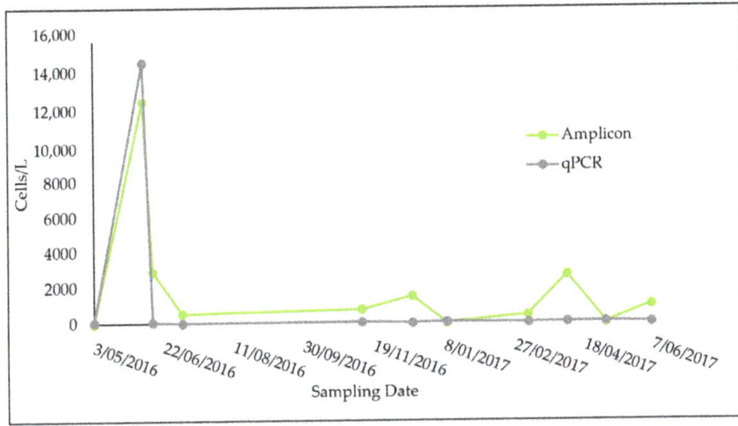

Figure 4. Bare Island *P. minimum* cell counts L^{-1} with amplicon sequencing and qPCR across the BPA sampling period. Standard error bars are included for qPCR (too small to detect).

The estimates of the cell abundances of *P. minimum* using all three methods showed results that were of the same order of magnitude and often relatively similar (Figures 3 and 4). The estimates of the abundance of *P. minimum* based on amplicon sequencing appeared to be consistently higher when compared with the qPCR and microscope count data (Figure 3). The highest abundances of *P. minimum* were found in June 2016; however, no microscope counts were completed for this month, so these data points were excluded in the following comparisons (Figures 3 and 4).

A highly significant relationship ($p = 1.61 \times 10^{-14}/p = 0.00$, Table 6) was found between the *P. minimum* abundance estimates derived from amplicon sequencing data and the *P. minimum*-specific qPCR, while the relationships between microscope counts and qPCR and microscope counts and amplicon sequencing were not significant ($p > 0.05$, Table 6). It is likely that the higher standard deviation in the method used for the light microscope cell count (Figure 3) may have led to the apparent differences in the cell counts of *P. minimum* using this method compared with that of the two molecular genetic methods.

Table 6. Linear regression between qPCR, metabarcoding, and light microscopy.

	qPCR vs. Amplicon Sequencing	qPCR vs. Count	Amplicon Sequencing vs. Count
Multiple R	0.90	0.18	0.23
R^2	0.82	0.03	0.05
Adjusted R^2	0.81	−0.02	0.00
Standard Error	1172.31	670.05	662.94
df	35	20	20
p (Significance)	0.00	0.43	0.31

3.4. Amplicon Sequencing Results

Using the amplicon sequencing method, 644 different phytoplankton ASVs in 428 genera were identified from the 27 samples from Towra Point, and 623 ASVs in 419 genera were identified from the 17 samples from Bare Island (Figure 5).

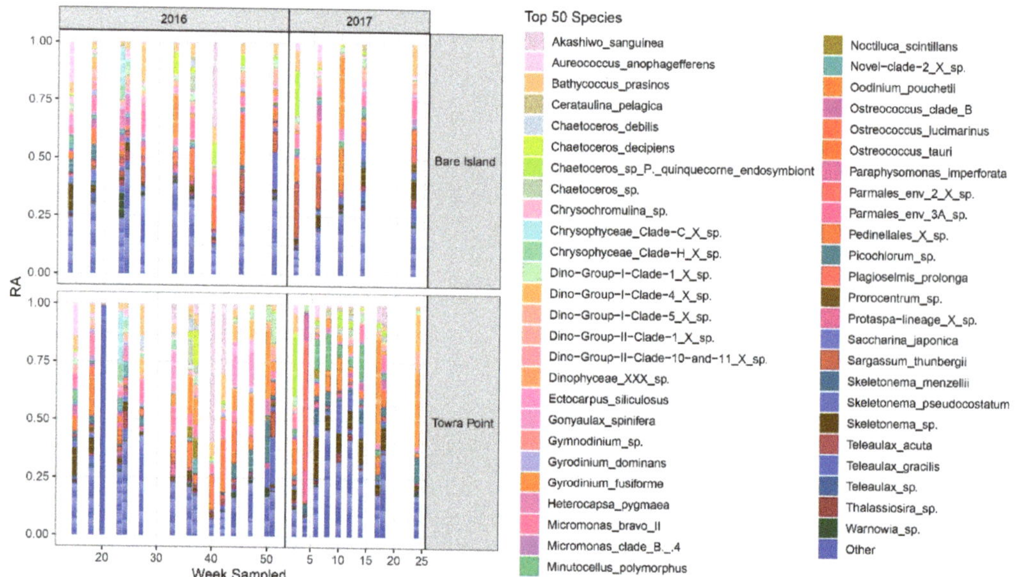

Figure 5. Relative abundance of the top 50 phytoplankton spp. at the Bare Island and Towra Point time series stations from April 2016 to June 2017. ASVs were assigned to phytoplankton species in the PR2.12 database. For clarity, only the top 50 by ASV abundance are shown. The less abundant taxa were lumped into the category "Other," while those without taxa assignments to the genus level were omitted.

3.5. Factors Influencing the Growth of P. minimum in Botany Bay

The abundance of *P. minimum* in Botany Bay was found to be significantly correlated to the environmental variables, total dissolved CO_2 (R = 0.34, p = 0.008), and salinity (R = −0.28, p = 0.035). All other variables were not found to be significantly correlated to *P. minimum* abundance. Data are submitted as Supplementary Material, S1: Physicochemical Data.

Co-occurrence Analysis

At Bare Island, *P. minimum* was found to have positive significant co-occurrence (p < 0.05) with four other phytoplankton species, *Pyramimonas gelidicola*, *Alexandrium pacificum*, *Euglyphida* sp., and *Goniomondales* sp. At Towra Point, *P. minimum* was found to have significant negative co-occurrence (p < 0.05) with two other phytoplankton species, *Prymnesiophyceae Clade F* sp. and *Blidingia dawsonii*, and positive significant associations (p < 0.05) with *Psammodictyon* sp., *Surirella* sp., *Tryblionella apiculata*, *Vampyrellida* sp., *Actinocyclus curvatulus*, *Cryothecomonas* sp., *Dino-Group-II-Clade-26* sp., *Massiteriidae* sp., *Navicula cryptocephala*, and *Navicula gregaria*.

4. Discussion

Prorocentrum minimum is a marine dinoflagellate that commonly occurs throughout the world and can form HABs [1,7,10]. HABs due to this species often occur in estuarine and coastal waters where aquaculture takes place, and in relation to that, death of shellfish has been reported [1,10]. *P. minimum* has been reported to show physiological flexibility with a global distribution across a range of conditions from temperate to subtropical [1,7,67]. It has been reported to produce TTX, a harmful neurotoxin [19,68–70]. Due to the potential harmful impacts of *P. minimum* on shellfish aquaculture, in this study, we aimed to develop new methods of investigating this species and apply them to environmental samples. In this study, a new culture of *P. minimum* was successfully isolated from Berowra Creek, Hawkesbury River, Australia. *P. minimum* has been linked to the production of TTX after a bloom in Vistonikos Bay, Greece, was positively correlated with TTX [27]. TTX was not detected in our strain. Genetic variability among strains may influence the toxicity of *P. minimum*, as well as the environmental conditions under which it is grown [1,24,26]. Due to the reported variability in toxicity in this species, future studies will be required to evaluate the toxicity of strains of *P. minimum*. As the alga is now successfully in culture with the Cawthron Institute Culture Collection, it allows future studies to look at more in depth toxin profiles, including how different environmental stressors and relationships with other known toxic algae or bacteria influence its toxicity.

qPCR assays have been developed over the past ~15 years for the detection and monitoring of HAB species [51,71,72]. qPCR has advantages over traditional light microscopy methods in that it is sensitive and rapid and allows for possible future automation. In the development of qPCR assays for the detection and quantification of specific taxa in environmental DNA, the most important considerations are the specificity of the assay in that it amplifies only the species of interest, and the amplification efficiency of assays with an efficiency of less than 90% will not give quantitative results across its full detection range [53,73]. Assays with an efficiency greater than 110% are considered to show significant inhibition to PCR amplification [74]. Amplification greater than 100% can be due to contamination in the sample, pipetting errors, inaccurate dilution series, and primer dimers [29]. For this study, a previously published qPCR assay developed for *P. minimum* was originally tested [51], which targeted a fragment of the small subunit ribosomal (SSU/18S) RNA gene. However, it was found to amplify several other nontarget *Prorocentrum* spp. and have a low efficiency (Table 5, 70%). Therefore, new primer sets were designed to develop a new qPCR assay for *P. minimum* with the aim of being specific, sensitive, and efficient. Twenty-two unique primer sets were designed and tested with variable results (Table 5). Only one of the primer sets was found to be sufficiently specific and efficient and was used to examine environmental samples for the presence of *P. minimum* (primer set 20, Table 5). The newly designed assay was based on the ITS rDNA gene region, which is more variable and faster evolving than the SSU rDNA gene among dinoflagellate species [75,76]. The assay did have a low level of cross-reactivity with the most genetically similar species, *P.* cf. *balticum*. However, *P.* cf. *balticum* could be distinguished from *P. minimum* due to a higher temperature on the melt curve profile. Several studies have used melt curve differences to discriminate similar species [77,78]. When analysed for efficiency, the new primer set showed E = 101% (*P. minimum* DNA) and E = 99% (gBlock synthetic DNA) (Table 5). The new assay amplifies a much shorter fragment than the previously published assay (71 bp compared with 325 bp), and this may account for its greater amplification efficiency [79]. The qPCR assay developed for *P. minimum* is more sensitive than most light microscopy counting methods, with a reliable detection limit of 13 cells L^{-1} [65,66].

Molecular barcoding, or amplicon sequencing, which involves PCR amplification of environmental DNA (eDNA) and then sequencing of short (~600) [80] "barcoding" gene regions using high-throughput sequencing (HTS), is another molecular genetic method that has begun to be used in phytoplankton research [81–83]. Amplicon sequencing uses broad-range primers designed to amplify conserved regions across whole domains of life—in this case, eukaryotes [56,84]. A major problem with the use of amplicon sequencing as a tool for quantifying microbial eukaryotes is that the "barcoding" genes may be present in multiple copies that can be variable among microalgal populations and species, meaning that sequenced gene amplicons may not reflect the true abundance of a species in the sample. qPCR is not immune to this problem; however, the impact is minimised by designing primers that amplify gene regions only present in a specific species and using a standard curve with known amounts of target.

However, in this study, the sequencing of amplicons of eukaryotic V4 regions of SSU rDNA from samples from Botany Bay did not show a significantly different quantity of *P. minimum* compared with the quantification based on qPCR (Figures 3 and 4 and Table 6). In addition, the results of this method have shown a previously unknown level of phytoplankton diversity in Botany Bay, detecting over 600 eukaryotic microbial ASVs between the two sites in Botany Bay. Previously, phytoplankton identification using light microscopy had led to the detection of only ~100 species or fewer in 10 years of phytoplankton monitoring at Botany Bay [85,86]. In this study, only 43% of all phytoplankton ASVs were able to be classified to species level using the 18S V4 primer set and the PR2 database [59]. Further development of reference databases of 18S V4 sequences from reference "voucher" specimen taxa curated by taxonomists is an important factor in the future of HTS to enable a more complete and accurate picture of microbiome species composition [87,88]. Another possible approach that may lead to a more specific identification of taxa is the use of other primer pairs that amplify other amplicon regions, such as the LSU rDNA region in dinoflagellates, the SSU (16S) plastid genes, CO1, cytochrome *b*, or other mitochondrial gene regions [82,83,89,90]. In previous studies, it was found that some groups of taxa, such as dinoflagellates, can be identified more readily using LSU rDNA regions, rather than SSU rDNA [83].

The collection and preservation of water samples for the identification and manual counting of cells with light microscopy has been the "gold standard" method used to study phytoplankton abundances [91–93]. The accuracy of light microscope-based microalgal enumeration is highly variable depending on the particular technique chosen, the counting effort, and the taxonomic expertise of the technician [81]. Compared with light microscopy enumeration, amplicon sequencing has been shown to be extremely sensitive and has the capacity to identify all phytoplankton species in a sample without requiring any taxonomic expertise. qPCR is an optimal technique for the enumeration of a particular target species, as the limit of detection is low, and the accuracy of the method is independent of the effort or taxonomic skills of the operator. It is relatively inexpensive, rapid, sensitive, and specific and, therefore, is highly suited to adoption for ongoing monitoring programs. qPCR can also be completed in situ at the time of sampling to get rapid results and can be used by trained shellfish producers to get results of HABs on-site.

The light microscopy counting method utilised in this study had a larger-than-average error rate, a high detection threshold of *P. minimum* (500 cells L^{-1} compared with 13 cells L^{-1} with qPCR), and comparatively fewer data points when compared with the molecular methods. Adoption of other light microscope counting methods, like the Utermöhl counting chamber [94], and the use of a DNA-based stain (i.e., a fluorescence in situ hybridisation (FISH) probe [95]) may have led to more accurate assessments of the abundance of *P. minimum* and the detection of cryptic species. For research into HAB ecology, a combination of the use of amplicon sequencing, to first determine the diversity of phytoplankton, particularly cryptic and small species, and then qPCR, to quantify the exact cell abundance, would appear to give optimal information for ecological inference and understanding of co-occurrence patterns. This two-step molecular pathway appears to be the most appropriate method for future development [29,35,71,92,96].

Botany Bay is an estuary in southeast Australia that is extensively modified [43], containing an international shipping port (Port Botany), an oil fuelling station, recreational beaches, industrial estates, and urban developments [97]. The bay is a highly populated area and is impacted by freshwater flows from the Georges and Cooks Rivers, both also extensively modified and surrounded with urban developments [97]. Despite the modification, Botany Bay is also home to a Ramsar wetland and one remaining oyster farm, both at Towra Point [98]. Thus, it is an important site for ongoing monitoring of HAB species, as they can impact not only the shellfish production but also the quality of the water for recreational and industrial purposes. Botany Bay and the Georges River have both previously experienced HABs, including *Noctiluca scintillans*, *Alexandrium pacificum*, other *Alexandrium* spp., and *Heterocapsa* spp. [99].

Two sites in Botany Bay were sampled from April 2016 to June 2017: Bare Island and Towra Point (Figure 1). Due to the extensively modified nature of the bay and its surroundings, and the nutrient input that can occur in relation to this land runoff, it was expected that *P. minimum* may be abundant at these sites. It was also expected that *P. minimum* may be particularly high in abundance at Towra Point, which is impacted by freshwater flows, as this species has been shown to flourish in low salinities with high nutrient freshwater inputs [1,70]. *P. minimum* was found to be in low abundance for most of the sampling period at both sites, detected at ~30 cells L^{-1} at both sites for most of the year of sampling (Figures 3 and 4). There was one peak in the abundance of *P. minimum* (8000–14,000 cells L^{-1}) at both sites, on 7 June 2016 (Figures 3 and 4). This could still be considered a low value for *P. minimum*, which has been detected at "bloom" levels upwards of 10 million cells^{-L} [11]. The low presence of *P. minimum* is an important current baseline for monitoring the health of Botany Bay and other southeast Australian estuaries.

The abundance of *P. minimum* was found to have a significant positive relationship with total CO_2, contrary to a previous finding that found that increased CO_2 had no relationship with the abundance of *P. minimum* [100]. *P. minimum* was also found to have a weak but significant negative relationship with salinity, which supports previous findings that *P. minimum* grows preferentially in decreasing salinities [1,70,101]. *P. minimum* was not found to have a significant relationship with any other environmental variables; however, it is likely that there may be a time lag between an environmental change and increase or decrease in *P. minimum* [102,103]. Incorporating a measure of exposure of *P. minimum* to environmental variables would require a higher temporal sampling frequency than what was undertaken in the present study. However, the correlations we found (+ve CO2 and −ve salinity) between *P. minimum* and the environmental variables measured are hypothesis forming and should be further investigated in Australian waters.

An analysis of phytoplankton species that significantly co-occurred with *P. minimum* in Botany Bay is useful, as in the past, toxicity attributed to blooms of *P. minimum* may have been also associated with the presence of *Dinophysis* spp., which are the main causative agents of diarrhetic shellfish poisoning (DSP), even when present in low abundances, such as ~100 cells L^{-1} [1,21,22,104]. Due to the potential uncertainties of amplicon sequencing-based estimates of the absolute abundance of cells in a sample, the data were analysed as a presence/absence matrix across all sampling dates [29,37]. *P. minimum* at Towra Point was found to significantly co-occur with 12 other phytoplankton species and at Bare Island with 4 other phytoplankton species. Of all the co-occurring species, only 1 is a toxin-producing species, *Alexandrium pacificum*. *A. pacificum* is an important HAB species due to the severity of bloom impacts in Australia, New Zealand, Korea, Japan, and other countries [72,105–107]. *A. pacificum* produces paralytic shellfish toxins (PSTs). *P. minimum* blooms have previously been associated with symptoms characteristic of PSTs [68,69]; however, there is still a possibility that it can produce other toxins not yet classified [19,20].

5. Conclusions

A sensitive, specific, and efficient *P. minimum* qPCR assay was successfully developed and will allow for high-throughput information to be collected on the distribution and abundance of this species. A new strain of *P. minimum* was also isolated from Berowra Creek, NSW, and shown to not produce tetrodotoxin. *P. minimum* was found to generally be in a low abundance in Botany Bay across all seasons during the BPA Marine Microbes sampling period (April 2016–June 2017), with one peak in its abundance at Towra Point and Bare Island in June. *P. minimum* was found to be significantly correlated to total CO_2 and to a decrease in salinity at the sites in Botany Bay. Further field and laboratory studies may be useful to determine more detailed information on the environmental variables associated with blooms of *P. minimum* in southeast Australia. *P. minimum* was found to positively co-occur with *A. pacificum*, which produces PSTs. This association may be relevant to the management of harmful algal blooms in Botany Bay and other oyster-producing estuaries in southeast Australia. qPCR is a useful method for the monitoring of particular HAB species as it is rapid, specific, sensitive, and efficient, while the use of amplicon sequencing based on the V4 region of the 18S rDNA found a level of microbial eukaryotic species diversity that was approximately six times greater than that previously known from this site. In the future, these two methods may be combined as a valuable tool for HAB research in Australian waters.

Supplementary Materials: The following are available online at https://www.mdpi.com/2076-2607/9/3/510/s1, Table S1: Physicochemical data, Table S2: Towra Point ASV data, Table S3: Bare Island ASV data.

Author Contributions: Conceptualisation, S.M.; methodology, K.M. and M.O.; software, K.M. and M.O.; validation, K.M., R.R., and S.M.; formal analysis, K.M. and M.O.; investigation, K.M.; resources, K.M.; data curation, K.M. and M.O.; writing—original draft preparation, K.M.; writing—review and editing, S.M., R.R., and M.O.; visualisation, S.M., K.M., and M.O.; supervision, S.M. and R.R.; project administration, R.R.; funding acquisition, S.M. All authors have read and agreed to the published version of the manuscript.

Funding: This research received no external funding.

Institutional Review Board Statement: Not applicable.

Informed Consent Statement: Not applicable.

Data Availability Statement: The data presented in this study are available in Supplementary Material; Table S1 for physicochemical data, Table S2 for Towra Point ASVs and Table S3 for Bare Island ASVs.

Acknowledgments: We would like to acknowledge the contribution of the Coastal and Benthic Marine Microbes consortium in the generation of data used in this publication. The Coastal and Benthic Marine Microbes project was supported by funding from Bioplatforms Australia and the Integrated Marine Observing System (IMOS) through the Australian government's National Collaborative Research Infrastructure Strategy (NCRIS) in partnership with the Australian research community. We thank Arjun Verma for assistance with lab work. We thank Abanti Barua for assistance with *P. minimum* culture isolation. We thank Penelope Ajani for her guidance on *P. minimum* in Australia, as well as with statistical analyses. We thank Chowdhury Sarowar at the Sydney Institute of Marine Science (SIMS) marine biotoxin facility for the toxin analysis and Steve Brett (Microalgal Services), Hazel Farrell, and Anthony Zammit (NSW Food Authority) for the Towra Point *P. minimum* microscopy counts. We thank the Food Agility CRC NSW Oysters Transformation Project for the support with sample collection in Botany Bay. We thank Ana Rubio and her team at Hornsby Shire Council for the collection of water samples containing *P. minimum* cells.

Conflicts of Interest: The authors declare no conflict of interest.

References

1. Heil, C.A.; Glibert, P.M.; Fan, C. Prorocentrum minimum (Pavillard) Schiller: A review of a harmful algal bloom species of growing worldwide importance. *Harmful Algae* **2005**, *4*, 449–470. [CrossRef]
2. Gobler, C.J.; Doherty, O.M.; Hattenrath-Lehmann, T.K.; Griffith, A.W.; Kang, Y.; Litaker, R.W. Ocean warming since 1982 has expanded the niche of toxic algal blooms in the North Atlantic and North Pacific oceans. *Proc. Natl. Acad. Sci. USA* **2017**, *114*, 4975–4980. [CrossRef]
3. Nixon, S.W. Coastal marine eutrophication: A definition, social causes, and future concerns. *Ophelia* **1995**, *41*, 199–219. [CrossRef]
4. Cloern, J.E. Our evolving conceptual model of the coastal eutrophication problem. *Mar. Ecol. Prog. Ser.* **2001**, *210*, 223–253. [CrossRef]
5. Ménesguen, A.; Lacroix, G. Modelling the marine eutrophication: A review. *Sci. Total Environ.* **2018**, *636*, 339–354. [CrossRef] [PubMed]
6. *FAO The State of World Fisheries and Aquaculture: Meeting the Sustainable Development Goals*; FAO: Rome, Italy, 2018.
7. Glibert, P.M.; Mayorga, E.; Seitzinger, S. Prorocentrum minimum tracks anthropogenic nitrogen and phosphorus inputs on a global basis: Application of spatially explicit nutrient export models. *Harmful Algae* **2008**, *8*, 33–38. [CrossRef]
8. Hajdu, S.; Pertola, S.; Kuosa, H. Prorocentrum minimum (Dinophyceae) in the Baltic Sea: Morphology, occurrence—A review. *Harmful Algae* **2005**, *4*, 471–480. [CrossRef]
9. Pertola, S.; Kuosa, H.; Olsonen, R. Is the invasion of Prorocentrum minimum (Dinophyceae) related to the nitrogen enrichment of the Baltic Sea? *Harmful Algae* **2005**, *4*, 481–492. [CrossRef]
10. Skarlato, S.; Telesh, I.; Mantnaseva, O.; Pozdnyakov, I.; Berdieva, M.; Schubert, H.; Filatova, N.; Knyazev, N.; Pechkovskaya, S. Studies of bloom-forming dinoflagellates Prorocentrum minimum in fluctuating environment: Contribution to aquatic ecology, cell biology and invasion theory. *Protistology* **2018**, *12*, 113–157. [CrossRef]
11. Ajani, P.A.; Larsson, M.E.; Woodcock, S.; Rubio, A.; Farrell, H.; Brett, S.; Murray, S.A. Bloom drivers of the potentially harmful dinoflagellate Prorocentrum minimum (Pavillard) Schiller in a south eastern temperate Australian estuary. *Estuar. Coast. Shelf Sci.* **2018**, *215*, 161–171. [CrossRef]
12. Heisler, J.; Glibert, P.; Burkholder, J.; Anderson, D.; Cochlan, W.; Dennison, W.; Gobler, C.; Dortch, Q.; Heil, C.; Humphries, E.; et al. Eutrophication and Harmful Algal Blooms: A Scientific Consensus. *Harmful Algae* **2008**, *8*, 3–13. [CrossRef]
13. Jeong, B.; Jeong, E.-S.; Malazarte, J.M.; Sin, Y. Physiological and Molecular Response of Prorocentrum minimum to Tannic Acid: An Experimental Study to Evaluate the Feasibility of Using Tannic Acid in Controling the Red Tide in a Eutrophic Coastal Water. *Int. J. Environ. Res. Public Health* **2016**, *13*, 503. [CrossRef] [PubMed]
14. Hajdu, S.; Edler, L.; Olenina, I.; Witek, B. Spreading and Establishment of the Potentially Toxic DinoflagellateProrocentrum minimum in the Baltic Sea. *Int. Rev. Hydrobiol.* **2000**, *85*, 561–575. [CrossRef]
15. Telesh, I.; Schubert, H.; Skarlato, S.O. Ecological niche partitioning of the invasive dinoflagellate Prorocentrum minimum and its native congeners in the Baltic Sea. *Harmful Algae* **2016**, *59*, 100–111. [CrossRef] [PubMed]
16. Tango, P.J.; Magnien, R.; Butler, W.; Luckett, C.; Luckenbach, M.; Lacouture, R.; Poukish, C. Impacts and potential effects due to Prorocentrum minimum blooms in Chesapeake Bay. *Harmful Algae* **2005**, *4*, 525–531. [CrossRef]
17. Tyler, M.A.; Seliger, H.H. Selection for a red tide organism: Physiological responses to the physical environment1,2. *Limnol. Oceanogr.* **1981**, *26*, 310–324. [CrossRef]
18. Stoecker, D.; Li, A.; Coats, D.; Gustafson, D.; Nannen, M. Mixotrophy in the dinoflagellate Prorocentrum minimum. *Mar. Ecol. Prog. Ser.* **1997**, *152*, 1–12. [CrossRef]
19. Denardou-Queneherve, A.; Grzebyk, D.; Pouchus, Y.; Sauviat, M.; Alliot, E.; Biard, J.; Berland, B.; Verbist, J. Toxicity of French strains of the dinoflagellate Prorocentrum minimum experimental and natural contaminations of mussels. *Toxicon* **1999**, *37*, 1711–1719. [CrossRef]
20. Grzebyk, D.; Denardou, A.; Berland, B.; Pouchus, Y.F. Evidence of a new toxin in the red-tide dinoflagellate *Prorocentrum minimum*. *J. Plankton Res.* **1997**, *19*, 1111–1124. [CrossRef]
21. Langeland, G.; Hasselgård, T.; Tangen, K.; Skulberg, O.M.; Hjelle, A. An outbreak of paralytic shellfish poisoning in western Norway. *Sarsia* **1984**, *69*, 185–193. [CrossRef]
22. Tangen, K. Shellfish poisoning and the occurrence of potentially toxic dinoflagellates in Norwegian waters. *Sarsia* **1983**, *68*, 1–7. [CrossRef]
23. Landsberg, J.H. The Effects of Harmful Algal Blooms on Aquatic Organisms. *Rev. Fish. Sci.* **2002**, *10*, 113–390. [CrossRef]
24. Wikfors, G. A review and new analysis of trophic interactions between Prorocentrum minimum and clams, scallops, and oysters. *Harmful Algae* **2005**, *4*, 585–592. [CrossRef]

25. Ogburn, D.; Callinan, R.; Pearce, I.; Hallegraeff, G.; Landos, M. Investigation and Management of a Major Oyster Mortality Event in Wonboyn Lake, Australia. In *Diseases in Asian Aquaculture*; Walker, P., Lester, R., Bondad-Reantaso, M.G., Eds.; Asian Fisheries Society: Manila, Philippines, 2005; pp. 301–309.

26. Rodríguez, I.; Alfonso, A.; Alonso, E.; Rubiolo, J.A.; Roel, M.; Vlamis, A.; Katikou, P.; Jackson, S.A.; Menon, M.L.; Dobson, A.; et al. The association of bacterial C9-based TTX-like compounds with Prorocentrum minimum opens new uncertainties about shellfish seafood safety. *Sci. Rep.* **2017**, *7*, 40880. [CrossRef]

27. Vlamis, A.; Katikou, P.; Rodriguez, I.; Rey, V.; Alfonso, A.; Papazachariou, A.; Zacharaki, T.; Botana, A.; Botana, L.; Vlamis, A.; et al. First Detection of Tetrodotoxin in Greek Shellfish by UPLC-MS/MS Potentially Linked to the Presence of the Dinoflagellate Prorocentrum minimum. *Toxins* **2015**, *7*, 1779–1807. [CrossRef]

28. Park, B.S.; Guo, R.; Lim, W.-A.; Ki, J.-S. Importance of free-living and particle-associated bacteria for the growth of the harmful dinoflagellate Prorocentrum minimum: Evidence in culture stages. *Mar. Freshw. Res.* **2018**, *69*, 290. [CrossRef]

29. Mäki, A.; Salmi, P.; Mikkonen, A.; Kremp, A.; Tiirola, M. Sample Preservation, DNA or RNA Extraction and Data Analysis for High-Throughput Phytoplankton Community Sequencing. *Front. Microbiol.* **2017**, *8*, 1848. [CrossRef]

30. Sellner, K.G.; Doucette, G.J.; Kirkpatrick, G.J. Harmful algal blooms: Causes, impacts and detection. *J. Ind. Microbiol. Biotechnol.* **2003**, *30*, 383–406. [CrossRef]

31. Medlin, L. Molecular tools for monitoring harmful algal blooms. *Environ. Sci. Pollut. Res.* **2013**, *20*, 6683–6685. [CrossRef]

32. Lefterova, M.I.; Budvytiene, I.; Sandlund, J.; Färnert, A.; Banaei, N. Simple Real-Time PCR and Amplicon Sequencing Method for Identification of Plasmodium Species in Human Whole Blood. *J. Clin. Microbiol.* **2015**, *53*, 2251–2257. [CrossRef]

33. Medlin, L.; Orozco, J. Molecular Techniques for the Detection of Organisms in Aquatic Environments, with Emphasis on Harmful Algal Bloom Species. *Sensors* **2017**, *17*, 1184. [CrossRef] [PubMed]

34. Kudela, R.M.; Howard, M.D.A.; Jenkins, B.D.; Miller, P.E.; Smith, G.J. Using the molecular toolbox to compare harmful algal blooms in upwelling systems. *Prog. Oceanogr.* **2010**, *85*, 108–121. [CrossRef]

35. Murray, D.C.; Coghlan, M.L.; Bunce, M. From Benchtop to Desktop: Important Considerations when Designing Amplicon Sequencing Workflows. *PLoS ONE* **2015**, *10*, e0124671. [CrossRef] [PubMed]

36. Stern, R.F.; Horak, A.; Andrew, R.L.; Coffroth, M.-A.; Andersen, R.A.; Küpper, F.C.; Jameson, I.; Hoppenrath, M.; Véron, B.; Kasai, F.; et al. Environmental barcoding reveals massive dinoflagellate diversity in marine environments. *PLoS ONE* **2010**, *5*, e13991. [CrossRef] [PubMed]

37. Galluzzi, L.; Bertozzini, E.; Penna, A.; Perini, F.; Garcés, E.; Magnani, M. Analysis of rRNA gene content in the Mediterranean dinoflagellate Alexandrium catenella and Alexandrium taylori: Implications for the quantitative real-time PCR-based monitoring methods. *J. Appl. Phycol.* **2010**, *22*, 1–9. [CrossRef]

38. Godhe, A.; Asplund, M.E.; Härnström, K.; Saravanan, V.; Tyagi, A.; Karunasagar, I. Quantification of diatom and dinoflagellate biomasses in coastal marine seawater samples by real-time PCR. *Appl. Environ. Microbiol.* **2008**, *74*, 7174–7182. [CrossRef]

39. Bachvaroff, T.R.; Place, A.R. From Stop to Start: Tandem Gene Arrangement, Copy Number and Trans-Splicing Sites in the Dinoflagellate Amphidinium carterae. *PLoS ONE* **2008**, *3*, e2929. [CrossRef]

40. Krehenwinkel, H.; Wolf, M.; Lim, J.Y.; Rominger, A.J.; Simison, W.B.; Gillespie, R.G. Estimating and mitigating amplification bias in qualitative and quantitative arthropod metabarcoding. *Sci. Rep.* **2017**, *7*, 1–12. [CrossRef]

41. Bradley, I.M.; Pinto, A.J.; Guest, J.S. Design and Evaluation of Illumina MiSeq-Compatible, 18S rRNA Gene-Specific Primers for Improved Characterization of Mixed Phototrophic Communities. *Appl. Environ. Microbiol.* **2016**, *82*, 5878–5891. [CrossRef]

42. Australian Microbiome–Australian Microbiome. Available online: https://www.australianmicrobiome.com/ (accessed on 9 December 2020).

43. Ajani, P.; Brett, S.; Krogh, M.; Scanes, P.; Webster, G.; Armand, L. The risk of harmful algal blooms (HABs) in the oyster-growing estuaries of New South Wales, Australia. *Environ. Monit. Assess.* **2013**, *185*, 5295–5316. [CrossRef]

44. Woelkerling, W.J.; Kowal, R.R.; Gough, S.B. Sedgwick-rafter cell counts: A procedural analysis. *Hydrobiologia* **1976**, *48*, 95–107. [CrossRef]

45. Throndsen, J. Preservation and storage. In *Phytoplankton Manual*; Sournia, A., Ed.; UNESCO: Paris, France, 1978; pp. 69–74. ISBN 9231015729.

46. Van de Kamp, J.; Mazard, S. Coastal Seawater Sampling for Australian Coastal Microbial Observatory Network. In *Australian Microbiome Methods*; Bioplatforms Australia: Sydney, Australia, 2020; pp. 12–18. Available online: https://www.australianmicrobiome.com/wp-content/uploads/2021/01/AM_Methods_for_metadata_fields_18012021_V1.2.3.pdf (accessed on 1 March 2019).

47. Keller, M.D.; Selvin, R.C.; Claus, W.; Guillard, R.R.L. Media for the culture of oceanic ultraphytoplankton. *J. Phycol.* **2007**, *23*, 633–638. [CrossRef]

48. Jeffrey, S.W.; LeRoi, J.M. Simple procedures for growing SCOR reference microalgal cultures. In *Phytoplankton Pigments in Oceanography: Monographs on Oceanographic Methodology*; Jeffrey, S.W., Mantoura, R.F.C., Wright, S.W., Eds.; UNESCO: Paris, France, 1997; pp. 181–205. ISBN 9231032755.

49. Harwood, T.; Boundy, M.; Selwood, A.; Ginkel, R.; MacKenzie, L.; McNabb, P. Refinement and implementation of the Lawrence method (AOAC 2005.06) in a commercial laboratory: Assay performance during an Alexandrium catenella bloom event. *Harmful Algae* **2013**, *24*, 20–31. [CrossRef]

50. Bio-Rad Laboratories Inc. T100TM Thermal Cycler. Available online: https://www.bio-rad.com/en-au/product/t100-thermal-cycler?ID=LGTWGIE8Z (accessed on 30 September 2019).

51. Handy, S.M.; Demir, E.; Hutchins, D.A.; Portune, K.J.; Whereat, E.B.; Hare, C.E.; Rose, J.M.; Warner, M.; Farestad, M.; Cary, S.C.; et al. Using quantitative real-time PCR to study competition and community dynamics among Delaware Inland Bays harmful algae in field and laboratory studies. *Harmful Algae* **2008**, *7*, 599–613. [CrossRef]

52. Bio-Rad Laboratories Inc. CFX96 Touch Real-Time PCR Detection System. Available online: https://www.bio-rad.com/en-au/product/cfx96-touch-real-time-pcr-detection-system?ID=LJB1YU15 (accessed on 9 September 2019).

53. Bustin, S.A.; Benes, V.; Garson, J.A.; Hellemans, J.; Huggett, J.; Kubista, M.; Mueller, R.; Nolan, T.; Pfaffl, M.W.; Shipley, G.L.; et al. The MIQE guidelines: Minimum information for publication of quantitative real-time PCR experiments. *Clin. Chem.* **2009**, *55*, 611–622. [CrossRef] [PubMed]

54. Taylor, S.C.; Nadeau, K.; Abbasi, M.; Lachance, C.; Nguyen, M.; Fenrich, J. The Ultimate qPCR Experiment: Producing Publication Quality, Reproducible Data the First Time. *Trends Biotechnol.* **2019**, *37*, 761–774. [CrossRef] [PubMed]

55. Larionov, A.; Krause, A.; Miller, W. A standard curve based method for relative real time PCR data processing. *BMC Bioinform.* **2005**, *6*, 62. [CrossRef]

56. Hadziavdic, K.; Lekang, K.; Lanzen, A.; Jonassen, I.; Thompson, E.M.; Troedsson, C. Characterization of the 18S rRNA gene for designing universal eukaryote specific primers. *PLoS ONE* **2014**, *9*, e87624. [CrossRef]

57. Bioplatforms Australia. *Protocol for 18S rRNA Amplification and Sequencing on the Illumina MiSeq*; Bioplatforms Australia: Sydney, Australia, 2015.

58. Amaral-Zettler, L.; Bauer, M.; Berg-Lyons, D.; Betley, J.; Caporaso, J.G.; Ducklow, H.W.; Fierer, N.; Fraser, L.; Gilbert, J.A.; Gormley, N.; et al. *EMP 18S Illumina Amplicon Protocol*.; Earth Microbiome Project. Available online: https://earthmicrobiome.org/protocols-and-standards/18s/ (accessed on 26 February 2021).

59. Guillou, L.; Bachar, D.; Audic, S.; Bass, D.; Berney, C.; Bittner, L.; Boutte, C.; Burgaud, G.; de Vargas, C.; Decelle, J.; et al. The Protist Ribosomal Reference database (PR2): A catalog of unicellular eukaryote Small Sub-Unit rRNA sequences with curated taxonomy. *Nucleic Acids Res.* **2012**, *41*, D597–D604. [CrossRef]

60. Callahan, B.J.; McMurdie, P.J.; Rosen, M.J.; Han, A.W.; Johnson, A.J.A.; Holmes, S.P. DADA2: High-resolution sample inference from Illumina amplicon data. *Nat. Methods* **2016**, *13*, 581–583. [CrossRef]

61. Croux, C.; Dehon, C. Influence functions of the Spearman and Kendall correlation measures. *Stat. Methods Appt.* **2010**, *19*, 497–515. [CrossRef]

62. Griffith, D.M.; Veech, J.A.; Marsh, C.J. cooccur: Probabilistic Species Co-Occurrence Analysis in R. *J. Stat. Softw.* **2016**, *69*, 1–17. [CrossRef]

63. Team, R.C. *R: A Language and Environment for Statistical Computing*; R Foundation: Vienna, Austria, 2019.

64. Veech, J.A. A probabilistic model for analysing species co-occurrence. *Glob. Ecol. Biogeogr.* **2013**, *22*, 252–260. [CrossRef]

65. Smith, C.J.; Nedwell, D.B.; Dong, L.F.; Osborn, A.M. Evaluation of quantitative polymerase chain reaction-based approaches for determining gene copy and gene transcript numbers in environmental samples. *Environ. Microbiol.* **2006**, *8*, 804–815. [CrossRef]

66. Smith, C.J.; Osborn, A.M. Advantages and limitations of quantitative PCR (Q-PCR)-based approaches in microbial ecology. *FEMS Microbiol. Ecol.* **2009**, *67*, 6–20. [CrossRef] [PubMed]

67. Zhu, X.; Zhen, Y.; Mi, T.; Yu, Z. Detection of Prorocentrum minimum (Pavillard) Schiller with an Electrochemiluminescence-Molecular Probe Assay. *Mar. Biotechnol. (NY)* **2012**. [CrossRef] [PubMed]

68. Silva, E.S.; Sousa, I. Experimental work on the dinoflagellate toxin production. *Arq. Inst. Nac. Saude* **1981**, *6*, 381–387.

69. Chen, Y.Q.; Gu, X. An ecological study of red tides in the East China Sea. In *Toxic Phytoplankton Blooms in the Sea*; Smayda, T.J., Shimizu, Y., Eds.; Elsevier: Amsterdam, The Netherlands, 1993; pp. 217–221.

70. Grzebyk, D.; Berland, B. Influences of temperature, salinity and irradiance on growth of *Prorocentrum minimum* (Dinophyceae) from the Mediterranean Sea. *J. Plankton Res.* **1996**, *18*, 1837–1849. [CrossRef]

71. Murray, D.C.; Bunce, M.; Cannell, B.L.; Oliver, R.; Houston, J.; White, N.E.; Barrero, R.A.; Bellgard, M.I.; Haile, J. DNA-Based Faecal Dietary Analysis: A Comparison of qPCR and High Throughput Sequencing Approaches. *PLoS ONE* **2011**, *6*, e25776. [CrossRef]

72. Ruvindy, R.; Bolch, C.J.; MacKenzie, L.; Smith, K.F.; Murray, S.A. qPCR Assays for the Detection and Quantification of Multiple Paralytic Shellfish Toxin-Producing Species of Alexandrium. *Front. Microbiol.* **2018**, *9*, 3153. [CrossRef]

73. Svec, D.; Tichopad, A.; Novosadova, V.; Pfaffl, M.W.; Kubista, M. How good is a PCR efficiency estimate: Recommendations for precise and robust qPCR efficiency assessments. *Biomol. Detect. Quantif.* **2015**, *3*, 9–16. [CrossRef] [PubMed]

74. Kontanis, E.J.; Reed, F.A. Evaluation of Real-Time PCR Amplification Efficiencies to Detect PCR Inhibitors. *J. Forensic Sci.* **2006**, *51*, 795–804. [CrossRef] [PubMed]

75. Murray, S.; Flø Jørgensen, M.; Ho, S.Y.W.; Patterson, D.J.; Jermiin, L.S. Improving the Analysis of Dinoflagellate Phylogeny based on rDNA. *Protist* **2005**, *156*, 269–286. [CrossRef] [PubMed]

76. Litaker, R.W.; Vandersea, M.W.; Kibler, S.R.; Reece, K.S.; Stokes, N.A.; Lutzoni, F.M.; Yonish, B.A.; West, M.A.; Black, M.N.D.; Tester, P.A. Recognizing dinoflagellate species using ITS rRNA sequences. *J. Phycol.* **2007**, *43*, 344–355. [CrossRef]

77. Andree, K.B.; Fernández-Tejedor, M.; Elandaloussi, L.M.; Quijano-Scheggia, S.; Sampedro, N.; Garcés, E.; Camp, J.; Diogène, J. Quantitative PCR coupled with melt curve analysis for detection of selected Pseudo-nitzschia spp. (Bacillariophyceae) from the northwestern Mediterranean Sea. *Appl. Environ. Microbiol.* **2011**, *77*, 1651–1659. [CrossRef] [PubMed]

78. Winder, L.; Phillips, C.; Richards, N.; Ochoa-Corona, F.; Hardwick, S.; Vink, C.J.; Goldson, S. Evaluation of DNA melting analysis as a tool for species identification. *Methods Ecol. Evol.* **2011**, *2*, 312–320. [CrossRef]
79. Bustin, S.; Huggett, J. qPCR primer design revisited. *Biomol. Detect. Quantif.* **2017**, *14*, 19–28. [CrossRef]
80. Kress, W.J.; García-Robledo, C.; Uriarte, M.; Erickson, D.L. DNA barcodes for ecology, evolution, and conservation. *Trends Ecol. Evol.* **2015**, *30*, 25–35. [CrossRef]
81. Godhe, A.; Cusack, C.; Pedersen, J.; Andersen, P.; Anderson, D.M.; Bresnan, E.; Cembella, A.; Dahl, E.; Diercks, S.; Elbrächter, M.; et al. Intercalibration of classical and molecular techniques for identification of Alexandrium fundyense (Dinophyceae) and estimation of cell densities. *Harmful Algae* **2007**, *6*, 56–72. [CrossRef]
82. Le Bescot, N.; Mahé, F.; Audic, S.; Dimier, C.; Garet, M.J.; Poulain, J.; Wincker, P.; de Vargas, C.; Siano, R. Global patterns of pelagic dinoflagellate diversity across protist size classes unveiled by metabarcoding. *Environ. Microbiol.* **2016**, *18*, 609–626. [CrossRef]
83. Smith, K.F.; Kohli, G.S.; Murray, S.A.; Rhodes, L.L. Assessment of the metabarcoding approach for community analysis of benthic-epiphytic dinoflagellates using mock communities. *New Zeal. J. Mar. Freshw. Res.* **2017**, *51*, 555–576. [CrossRef]
84. Hong, S.; Bunge, J.; Leslin, C.; Jeon, S.; Epstein, S.S. Polymerase chain reaction primers miss half of rRNA microbial diversity. *ISME J.* **2009**, *3*, 1365–1373. [CrossRef]
85. Ajani, P.A.; Hallegraeff, G.M.; Allen, D.; Coughlan, A.; Richardson, A.J.; Armand, L.K.; Ingleton, T.; Murray, S.A. Establishing Baselines: Eighty Years of Phytoplankton Diversity and Biomass in South-Eastern Australia. In *Oceanography and Marine Biology*; CRC Press: Boca Raton, FL, USA, 2016; pp. 395–420.
86. Ajani, P.A.; Allen, A.P.; Ingleton, T.; Armand, L.K. A decadal decline in relative abundance and a shift in microphytoplankton composition at a long-term coastal station off southeast Australia. *Limnol. Oceanogr.* **2014**, *59*, 519–531. [CrossRef]
87. Macheriotou, L.; Guilini, K.; Bezerra, T.N.; Tytgat, B.; Nguyen, D.T.; Phuong Nguyen, T.X.; Noppe, F.; Armenteros, M.; Boufahja, F.; Rigaux, A.; et al. Metabarcoding free-living marine nematodes using curated 18S and CO1 reference sequence databases for species-level taxonomic assignments. *Ecol. Evol.* **2019**, *9*, 1211–1226. [CrossRef] [PubMed]
88. Boers, S.A.; Jansen, R.; Hays, J.P. Understanding and overcoming the pitfalls and biases of next-generation sequencing (NGS) methods for use in the routine clinical microbiological diagnostic laboratory. *Eur. J. Clin. Microbiol. Infect. Dis.* **2019**, *38*, 1059–1070. [CrossRef] [PubMed]
89. Pochon, X.; Bott, N.J.; Smith, K.F.; Wood, S.A. Evaluating detection limits of next-generation sequencing for the surveillance and monitoring of international marine pests. *PLoS ONE* **2013**, *8*, e73935. [CrossRef]
90. Kohli, G.S.; Neilan, B.A.; Brown, M.V.; Hoppenrath, M.; Murray, S.A. Cob gene pyrosequencing enables characterization of benthic dinoflagellate diversity and biogeography. *Environ. Microbiol.* **2014**, *16*, 467–485. [CrossRef]
91. Galluzzi, L.; Penna, A.; Bertozzini, E.; Vila, M.; Garces, E.; Magnani, M. Development of a Real-Time PCR Assay for Rapid Detection and Quantification of Alexandrium minutum (a Dinoflagellate). *Appl. Environ. Microbiol.* **2004**, *70*, 1199–1206. [CrossRef]
92. Shaw, J.L.-A. *Metagenomic Amplicon Sequencing as a Rapid and High-Throughput Tool for Aquatic Biodiversity Surveys*; University of Adelaide: Adelaide, SA, Australia, 2015.
93. Penna, A.; Galluzzi, L. The quantitative real-time PCR applications in the monitoring of marine harmful algal bloom (HAB) species. *Environ. Sci. Pollut. Res. Int.* **2013**, *20*, 6851–6862. [CrossRef]
94. Paxinos, R. A rapid Utermohl method for estimating algal numbers. *J. Plankton Res.* **2000**, *22*, 2255–2262. [CrossRef]
95. Murray, S.A.; Ruvindy, R.; Kohli, G.S.; Anderson, D.M.; Brosnahan, M.L. Evaluation of sxtA and rDNA qPCR assays through monitoring of an inshore bloom of Alexandrium catenella Group 1. *Sci. Rep.* **2019**, *9*, 14532. [CrossRef]
96. Engesmo, A.; Strand, D.; Gran-Stadniczeñko, S.; Edvardsen, B.; Medlin, L.K.; Eikrem, W. Development of a qPCR assay to detect and quantify ichthyotoxic flagellates along the Norwegian coast, and the first Norwegian record of Fibrocapsa japonica (Raphidophyceae). *Harmful Algae* **2018**, *75*, 105–117. [CrossRef] [PubMed]
97. Loveless, A.M. A multi-dimensional receiving water quality model for Botany Bay (Sydney, Australia). In Proceedings of the 18th World IMACS/MODSIM Congress, Cairns, Australia, 13–17 July 2009; pp. 4170–4176.
98. DECCW. *Towra Point Nature Reserve Ramsar Site*; DECCW: Sydney, Australia, 2010.
99. Ajani, P.; Ingleton, T.; Pritchard, T.; Armand, L. Microalgal Blooms in the Coastal Waters of New South Wales, Australia. *Proc. Linn. Soc. New South. Wales* **2011**, *133*, 15–32.
100. Fu, F.X.; Zhang, Y.; Warner, M.E.; Feng, Y.; Sun, J.; Hutchins, D.A. A comparison of future increased CO2 and temperature effects on sympatric Heterosigma akashiwo and Prorocentrum minimum. *Harmful Algae* **2008**, *7*, 76–90. [CrossRef]
101. Skarlato, S.; Filatova, N.; Knyazev, N.; Berdieva, M.; Telesh, I. Salinity stress response of the invasive dinoflagellate Prorocentrum minimum. *Estuar. Coast. Shelf Sci.* **2018**, *211*, 199–207. [CrossRef]
102. Trombetta, T.; Vidussi, F.; Mas, S.; Parin, D.; Simier, M.; Mostajir, B. Water temperature drives phytoplankton blooms in coastal waters. *PLoS ONE* **2019**, *14*. [CrossRef] [PubMed]
103. Collos, Y. Time-lag algal growth dynamics: Biological constraints on primary production in aquatic environments. *Mar. Ecol. Prog. Ser.* **1986**, *33*, 193–206. [CrossRef]
104. Reguera, B.; Velo-Suárez, L.; Raine, R.; Park, M.G. Harmful Dinophysis species: A review. *Harmful Algae* **2012**, *14*, 87–106. [CrossRef]
105. Anderson, D.M.; Alpermann, T.J.; Cembella, A.D.; Collos, Y.; Masseret, E.; Montresor, M. The globally distributed genus Alexandrium: Multifaceted roles in marine ecosystems and impacts on human health. *Harmful Algae* **2012**, *14*, 10–35. [CrossRef]

106. Gao, Y.; Yu, R.C.; Chen, J.H.; Zhang, Q.C.; Kong, F.Z.; Zhou, M.J. Distribution of Alexandrium fundyense and A. pacificum (Dinophyceae) in the Yellow Sea and Bohai Sea. *Mar. Pollut. Bull.* **2015**, *96*, 210–219. [CrossRef]
107. Shin, H.H.; Li, Z.; Kim, E.S.; Park, J.-W.; Lim, W.A. Which species, Alexandrium catenella (Group I) or A. pacificum (Group IV), is really responsible for past paralytic shellfish poisoning outbreaks in Jinhae-Masan Bay, Korea? *Harmful Algae* **2017**, *68*, 31–39. [CrossRef] [PubMed]

Correction

Correction: McLennan et al. Assessing the Use of Molecular Barcoding and qPCR for Investigating the Ecology of *Prorocentrum minimum* (Dinophyceae), a Harmful Algal Species. *Microorganisms* 2021, 9, 510

Kate McLennan, Rendy Ruvindy, Martin Ostrowski and Shauna Murray *

Faculty of Science, University of Technology Sydney, Ultimo, NSW 2007, Australia
* Correspondence: shauna.murray@uts.edu.au

Citation: McLennan, K.; Ruvindy, R.; Ostrowski, M.; Murray, S. Correction: McLennan et al. Assessing the Use of Molecular Barcoding and qPCR for Investigating the Ecology of *Prorocentrum minimum* (Dinophyceae), a Harmful Algal Species. *Microorganisms* 2021, 9, 510. *Microorganisms* **2022**, *10*, 1906. https://doi.org/10.3390/microorganisms10101906

Received: 2 August 2022
Accepted: 22 August 2022
Published: 26 September 2022

Publisher's Note: MDPI stays neutral with regard to jurisdictional claims in published maps and institutional affiliations.

The authors wish to make the following corrections to this paper [1].
A correction has been made to Table 1:

Table 1. Names and sequences of all the primers used in this project. Primers 200F and 525R are taken from [51]. The other primers were designed using the online Primer-BLAST software (NCBI). All reverse primers are in reverse complement.

Name	Sequence (Forward)	Name	Sequence (Reverse)
		Primer Sequences for qPCR	
Pm 200F	TGTGTTTATTAGTTACAGAACCAGC	Pm 525R	AATTCTACTCATTCCAATTACAAGACAAT
1F Pmin	CGCAGCGAAGTGTGATAAGC	1R Pmin	TCTGGAAAGGCCAGAAGCTG
2F Pmin	TCGGCTCGAACAACGATGAA	2R Pmin	AAGCGTTCTGGAAAGGCCAG
3F Pmin	TTCTGGCCTTTCCAGAACGC	3R Pmin	CATGCCCAACAACAAGGCAA
4F Pmin	CGTATACTGCGCTTTCGGGA	4R Pmin	CACACAGAAACACACAAGCGT
5F Pmin	CCTTTCCAGAACGCTTGTGTG	5R Pmin	CTGGGCACTAGACAGCAAGG
6F Pmin	CAGGCTCAGACCGTCTTCTG	6R Pmin	AGCGTTCTGGAAAGGCCAG
7F Pmin	CAACAGTTGGTGAGGCTCT	7R Pmin	ATTCAAAAACACAGAAGATCAGGAA
8F Pmin	AACAACAGTTGGTGAGGCTCTG	8R Pmin	CAAAAACACAGAAGATCAGGAAGAC
9F Pmin	GTGAGGCTCTGGGTGGG	9R Pmin	CAAAAACACAGAAGATCAGGAAGAC
10F Pmin	TCATTCGCACGCATCCATTC	10R Pmin	AAGGACAGGCACAGAAGACG
11F Pmin	TTCAGTGCACAGGGTCTTCC	11R Pmin	GTCTTGGTAGGAGTGCGCTG
12F Pmin	GCCTTTCCAGAACGCTTGTGT	12R Pmin	GCTGACCTAACTTCATGTCTTGG
13F Pmin	CGCTTGTGTGTTTCTGTGTG	13R Pmin	CCATGCCCAACAACAAGGC
14F Pmin	TCTTCCCACGCAAGCAACT	14R Pmin	CGGGTTTGCTGACCTAAACT
15F Pmin	ACATTCGCACGCATCCATTC	15R Pmin	TTGCTGCCCTTGAGTCTCTG
16F Pmin	AACAGTTGGTGAGGCTCTGG	16R Pmin	AAGGACAGGCACAGAAGACG
17F Pmin	ACAACAGTTGGTGAGGCTCT	17R Pmin	TTGCTGCCCTTGAGTCTCTG
18F Pmin	CAGTTGGTGAGGCTCTGGG	18R Pmin	CAGAAGACGGTCTGAGCCTG
19F Pmin	TTCAGTGCACAGGGTCTTCC	19R Pmin	CATGCCCAACAACAAGGCAA
20F Pmin	ATTCCAGCTTCTGGCCTGTC	20R Pmin	TAGTTGCTTGCGTGGGAAGA
21F Pmin	CTGTCCAGAACGCTTGTGTG	21R Pmin	CTTCTAGTTGCTTGCGTGGG
22F Pmin	TTCCCACGCAAGCAACTAGA	22R Pmin	GCACTAGACAGCAAGGCCA
		Primer Sequences for Amplicon Sequencing	
Modified TAReuk454FWD1	CCAGCASCYGCGGTAATTCC	Modified TAReukREV3	ACTTTCGTTCTTGATYRATGA
		Primer Sequences for PCR and Sanger Sequencing	
d1f	ACCCGCTGAATTTAAGCATA	d3b	TCGGAGGGAACCAGCTACTA
ITSfwd	TTCCGTAGGTGAACCTGCGG	ITSrev	ATATGCTTAAATTCAGCGGGT

The authors apologize for any inconvenience caused and state that the scientific conclusions are unaffected. The original publication has also been updated.

Reference

1. McLennan, K.; Ruvindy, R.; Ostrowski, M.; Murray, S. Assessing the Use of Molecular Barcoding and qPCR for Investigating the Ecology of *Prorocentrum minimum* (Dinophyceae), a Harmful Algal Species. *Microorganisms* **2021**, *9*, 510. [CrossRef] [PubMed]

MDPI
St. Alban-Anlage 66
4052 Basel
Switzerland
Tel. +41 61 683 77 34
Fax +41 61 302 89 18
www.mdpi.com

Microorganisms Editorial Office
E-mail: microorganisms@mdpi.com
www.mdpi.com/journal/microorganisms